AIRCRAFT RECIPROCATING ENGINES

JEPPESEN
Sanderson Training Products

Library of Congress Cataloging-in-publication number: **92-24625**

JS312649B

Table of Contents

PREFACE

This book on *Aircraft Reciprocating Engines* is one of a series of specialized training manuals prepared for aviation maintenance personnel.

. This series is part of a programmed learning course developed and produced by Jeppesen Sanderson, Inc., one of the largest suppliers of aviation maintenance training materials in the world. This program is part of a continuing effort to improve the quality of education for aviation mechanics throughout the world.

The purpose of each training series is to provide basic information on the operation and principles of the various aircraft systems and their components. Specific information on detailed operation procedures should be obtained from the manufacturer's maintenance manuals, and followed in detail for best results.

This particular manual on *Aircraft Reciprocating Engines* includes a series of carefully prepared questions and answers to emphasize key elements of study, and to encourage you to continually test yourself for accuracy and retention as you use this book. A multiple choice final examination is included to allow you to test your comprehension of the total material.

Some of the words may be new to you. They are defined in the Glossary found at the back of the book.

Acknowledgments

The validity of any program such as this is enhanced immeasurably by the cooperation shown Jeppesen Sanderson by recognized experts in the field, and by the willingness of the various manufacturers to share their literature and answer countless questions. We are indebted to the many companies and individuals who provided assistance in the preparation of these training materials.

For product, service, or sales information call 1-800-621-JEPP, 303-799-9090, or FAX 303-784-4153. If you have comments, questions, or need explanations about any Maintenance Training System, we are prepared to offer assistance at any time. If your dealer does not have a Jeppesen catalog, please request one and we will promptly send it to you. Just call the above telephone number, or write:

Marketing Manager, Training Products
Jeppesen Sanderson, Inc.
55 Inverness Drive East
Englewood, CO 80112-5498

Please direct inquiries from Europe, Africa, and the Middle East to:

Jeppesen & Co., GmbH
P. O. Box 70-05-51
Walter-Kolb-Strasse 13
60594 Frankfurt
GERMANY
Tel: 011-49-69-961240
Fax: 011-49-69-96124898

Introduction

The concept of flying is as old as mythology, with one of the earliest stories of flying being that of Daedalus, an early-day craftsman and inventor who, along with his son Icarus, endeavored to escape from prison with wings of wax. The earliest actual flights were made in balloons and later in gliders. But, it was only when power became available that flight was actually practical.

Steam engines, compressed air, and even clockworks were experimented with to power some of the early flying contraptions. However, the only device that proved practical was the internal combustion gasoline engine.

Primitive automobile engines did not lend themselves to installation in the crude airplanes, so a new breed of engine had to be developed, engines in which low weight was of paramount concern.

When World War I came along, the rotary engine with its whirling cylinders gained popularity as the lightest engine for its power. This basic design evolved into the radial engine which held center stage for two decades.

Today, with turbine engines taking over the job of powering the large, high-speed aircraft, the field has been left open for the smooth, dependable, and relatively economical horizontally-opposed piston engine to power our fleet of general aviation aircraft.

While the turbine engine has gained popularity, it has not, nor will it ever, completely replace the reciprocating engine for aircraft propulsion. The piston engine is here to stay for small aircraft and for relatively short flights where the advantages of jet propulsion do not exist.

The aviation maintenance technician must understand the operating principles of the reciprocating engines he is to service, and, it is for this reason that this training manual has been prepared.

The generation of power in a reciprocating engine along with the factors which affect the amount of power the engine can develop are discussed, along with the construction and operation of the aircraft engine in its various configurations.

Engine servicing and overhaul are discussed to make the technician aware of the types of inspections and overhauls the engine must be given. And, finally, a brief overview of what is involved in a major overhaul of an aircraft engine is included with a short section on the basics of engine troubleshooting.

A basic principle permeates this book, that being, that any service performed on any aircraft engine must be done in strict accordance with the manufacturer's recommendations. This cannot be emphasized too strongly, as it is the manufacturer of the engine and its accessories who knows more about his product than any one else, and who is more concerned with his product performing with the utmost in safety and economy.

Chapter I

Theory And Classification

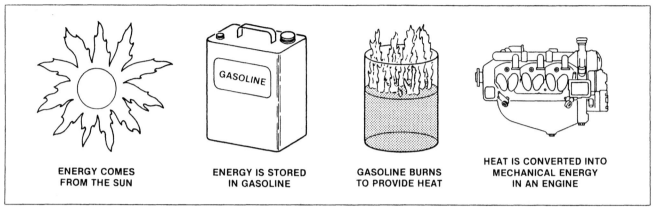

| ENERGY COMES FROM THE SUN | ENERGY IS STORED IN GASOLINE | GASOLINE BURNS TO PROVIDE HEAT | HEAT IS CONVERTED INTO MECHANICAL ENERGY IN AN ENGINE |

Figure 1-1. *The origin of all of our energy is the sun. It was received and used by animal and plant life which became our petroleum — the fuel for our aircraft engines.*

A. Basic Energy Transformation

1. Energy Conversion in Heat Engines

The internal combustion engine used in our modern aircraft is a form of heat engine; that is, it is a device which changes heat energy into mechanical energy. The energy used in the engine came, eons ago, from our chief source, the sun, and was changed into chemical energy by the photosynthesis of plant life. Then during some great upheaval in the past, the plants — and animals which ate them — were buried beneath thousands of tons of earth. Through heat and pressure, they became petroleum products as we know them today — fuel for our heat engines.

Petroleum is an organic chemical, a compound of hydrogen and carbon, and is of the family called hydrocarbons. Gasoline is refined from this crude petroleum to provide us with a usable form of chemical energy, and all that is needed then, is a device that will transform it into mechanical energy. This is where our heat engine comes into being.

Each pound of aviation gasoline has a heat energy content of approximately 20,000 British Thermal Units, with each Btu having enough heat energy to raise the temperature of one pound of water one degree Fahrenheit (59° to 60°).

The law of conservation of energy states that energy can be neither created nor destroyed, but can be changed in its form; and that is what is done when

gasoline, trapped in an engine cylinder, is ignited — the gasoline changes its chemical energy into heat energy. There is an abundance of energy in the fuel; the problem is in harnessing it.

2. Types of Heat Engines

a. External Combustion Engines

Heat engines fall basically into three groups. The first of these, the external combustion engine, was used for years in the reciprocating steam locomotive. A wood, coal, or oil fire was used to heat water until it turned into steam and expanded. The pressure from this expanding gas moved the pistons in the engine.

More modern external combustion engines use the steam to spin turbine wheels rather than moving the less efficient pistons.

Figure 1-2. *The external combustion steam engine produces power, but it is extremely inefficient.*

b. Rocket Engines

A second type of heat engine is the rocket, whose fuel, in either liquid or solid form, contains its own oxygen. Once the chemical reaction within the fuel is started, the internally-generated heat expands and accelerates the gases as they leave the engine, producing thrust independent of the atmosphere. This principle makes rockets usable in space craft. RATO (rocket-assisted take-off devices) are used for assisting heavily loaded aircraft on takeoff.

c. Internal Combustion Engines

The most popular form of heat engine used in aircraft today is the internal combustion engine. These engines may be divided into two classes: gas turbines and reciprocating engines. We will briefly describe each type here and go into the reciprocating engine in much more detail further on in this book.

(1) Gas Turbines

Gas turbine engines are ideal for flight because of their excellent power-to-weight ratio which is greater than four times that of a reciprocating engine. Air is taken into the engine through a carefully designed intake duct and then compressed by a large axial-flow or centrifugal compressor. When it leaves the compressor, the air enters a combustion chamber where fuel is sprayed into it and burned. The heat energy released when the burning fuel expands the air accelerates it as it leaves the engine, producing thrust. As these heated gases leave the combustion chambers, some of their energy is extracted to spin a turbine which drives the compressor.

Jet aircraft are propelled by the thrust produced when the gases passing through the engine are accelerated by the addition of the energy from the burning fuel. Turboprop and turboshaft engines extract a

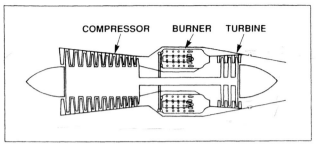

Figure 1-4. The gas turbine engine burns its fuel internally to release the heat needed to produce thrust.

great deal of energy from the expanding gases as they spin several stages of turbines which drive the propellers or rotors through reduction gears.

(2) Reciprocating Engines

There are two forms of reciprocating engines: the two-cycle which is difficult to cool and inefficient and is not used in any currently produced aircraft; and the four-cycle engine, which is used in almost all automobiles and in all modern aircraft.

(a) Two-stroke Cycle

The operating principle of a two-stroke cycle engine is shown in figure 1-6.

As the piston moves downward on the power stroke, just before it reaches its bottom center position, it passes a series of open exhaust ports, and the pressure within the cylinder forces the exhaust gases out. As these gases leave the cylinder, the single intake valve opens, and a fresh charge of fuel and air is brought in from the carburetor through the supercharger. The incoming charge aids in the

Figure 1-3. Rocket assisted takeoff (RATO) uses a solid propellant rocket engine to produce the extra thrust needed for getting out of tight places.

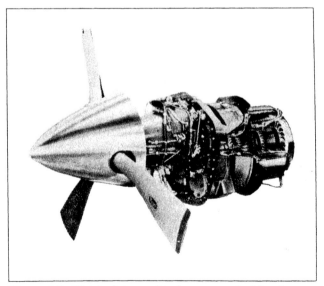

Figure 1-5. The turboprop engine uses a turbine to drive its gas generator, but produces thrust by spinning a propeller.

2

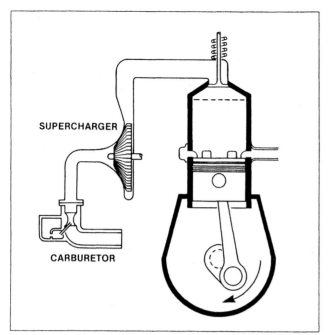

Figure 1-6. *The two-stroke cycle engine produces a power stroke every time the piston moves down, but its inefficiency and difficulty in cooling make it unacceptable as an aircraft engine.*

scavenging of the exhaust gases, and by the time the piston has reached the bottom and moved back up enough to cover the exhaust ports, the intake valve has closed. Further movement of the piston compresses the fuel-air mixture, and near the top of the stroke, the spark plug ignites the mixture so that the pressure within the cylinder can build up and force the piston down for the next cycle.

(b) Four-stroke Cycle

All aircraft engines are designed to operate according to the four-stroke cycle principle; that is, they produce one power stroke for each two complete revolutions of the crankshaft, or four strokes of the piston. A stroke is represented by the total movement of the piston from the bottom to the top center of the cylinder, or vice versa. This movement is equal to 180° of crankshaft rotation.

The four-stroke, five-event cycle was developed by August Otto in Germany about one hundred years ago, and the relationship between the pressure within the cylinder and the cylinder volume for this cycle is illustrated by the diagram of figure 1-8.

(1) Intake Stroke

The intake stroke actually begins as the piston leaves top center and starts down, but the intake valve has opened approximately 15° before the piston reached top center, to take advantage of the inertia of the

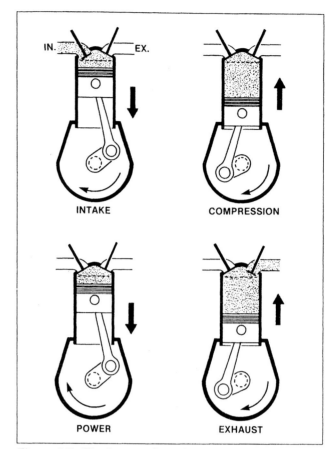

Figure 1-7. *The four strokes of a four-stroke cycle internal combustion engine.*

departing exhaust gases and get the maximum amount of fuel-air charge into the cylinders.

The intake valve remains open during the entire downward movement of the piston and even after it starts back up. At both the top and bottom of the stroke, the crankshaft turns several degrees before

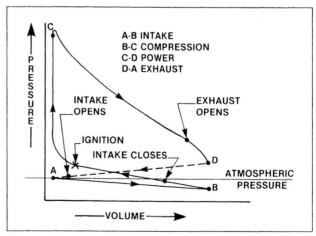

Figure 1-8. *Indicator diagram of the volume and pressure relationship of the Otto cycle of energy release.*

3

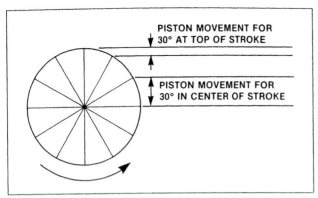

Figure 1-9. Piston movement vs. crankshaft travel.

the piston moves an appreciable amount as can be seen in the diagram of figure 1-9. This allows the intake valve to be left open until the crankshaft has rotated about 60° past bottom dead center, so the inertia of the fuel-air charge will get the maximum

amount into the cylinder. The timing of the opening and closing of the intake valve has a great deal to do with the volumetric efficiency of the engine. The more time the charge has to get into the cylinder, the higher will be the volumetric efficiency and the more power will be developed by the engine.

(2) Compression Stroke

The piston has started on its way back up before the intake valve closes, and now, with both valves closed, the fuel-air mixture is compressed as the piston continues to move toward top center. About 30° before it reaches top center, the spark plugs ignite the mixture, and the flame front moves across the head of the piston. Since the mixture burns rather than explodes, the piston passes slightly beyond top center and is on its way back down, when the maximum cylinder pressure is reached, and the piston gets its greatest push. If ignition occurs too early, the maximum pressure will

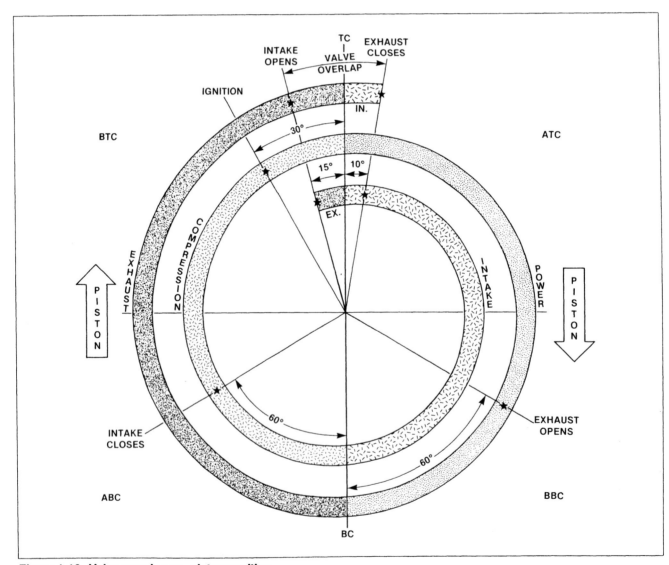

Figure 1-10. Valve opening vs. piston position.

4

be reached while the piston is moving upward, and power will be lost. If it occurs too late, the piston will be so far down on its stroke that the push will be ineffective and power will be lost.

(3) Power Stroke

The only time heat energy from the fuel is converted into actual work is during the power stroke. The heated and expanding gases force the piston down, performing work, and if time is taken into consideration, we can determine the amount of power produced.

The average pressure that actually exists in the cylinder during the power stroke is called the brake mean effective pressure (BMEP). This pressure, times the area of the piston head, gives the number of pounds of force, and when this is multiplied by the length of the stroke in feet, we have the foot-pounds of work done on each power stroke. If we know the number of power strokes in one minute we can compute the amount of power developed by the engine.

The exhaust valve opens about 55-60° of crankshaft rotation before the piston reaches the bottom of its power stroke, so the pressure inside the cylinder will start the exhaust gases moving out.

(4) Exhaust Stroke

The exhaust gases have started to be expelled from the cylinder by the time the piston reaches the bottom of its stroke, and as it moves back up, all of the burned gases are forced out. By the time the piston reaches the top of its stroke, the cylinder pressure will be low enough, and there will be enough inertia from the departing exhaust that the fuel-air charge will be assisted in entering the cylinder when the intake valve opens.

The timing of the opening and closing of the exhaust valves determines the efficiency of scavenging the gases, and anything that opposes the flow of gases from the cylinder will dilute the incoming fuel-air charge.

3. Work-power Considerations

a. Work

Work is accomplished when a force is moved through a specified distance, independent of time. In the English system of measurement, work is expressed in foot-pounds; in the metric system, in meter-kilograms. Work, measured in foot-pounds, is the product of the force in pounds times the distance through which the force acts, measured in feet. It may be found by the formula:

$$\text{Work} = \text{Force} \times \text{Distance}$$

If 1 lb. of mass is raised 1 ft., 1 ft.-lb. of work has been accomplished.

b. Power

Time is not involved in the determination of the amount of work done, but power does involve time. In fact, power may be defined as the time-rate of doing work, and is expressed in foot-pounds of work per minute, or per second.

$$\text{Power} = \text{Force} \times \frac{\text{Distance}}{\text{Time}}$$

c. Horsepower

James Watt, the inventor of the steam engine, found that an English dray horse could do, in a reasonable period of time, about 33,000 ft.-lbs. of work in one minute. And from his observations came the term "horsepower" which has been standardized as 33,000 ft.-lbs. of work done in one minute, or 550 ft.-lbs. per second. In electrical power there is a relationship between the watt and the horsepower. One horsepower is equal to 746 watts, or one kilowatt (1,000 watts) equals 1.34 HP.

When we work with aircraft reciprocating engines, there are several types of horsepower we must consider.

(1) Brake Horsepower

The actual amount of power delivered to the propeller shaft of an aircraft engine is called the brake horsepower (BHP) and derives its name from the method by which it is measured. In the early days of reciprocating engines, power was measured by clamping a brake around the output shaft and measuring the force exerted on an arm, at the speed the measurement was being taken. This measuring device was called a prony brake; thus the term, brake horsepower. Today the same type measurement may be made by the engine driving either a fluid pump or an electrical generator. The output of the pump or generator can be required to do work, and the power may be measured.

(2) Friction Horsepower

Pistons slide back and forth in the cylinders, and air is pulled into the engine and compressed. All of this requires power, as does the movement of the rotating machinery, the gears, and all of the accessories. All of the power actually required to drive the engine is lumped into one measurement called friction horsepower (FHP). This may be measured by driving the engine with a calibrated motor and finding the amount of power actually required to turn the engine at each speed.

(3) Indicated Horsepower

The power actually delivered to the propeller shaft plus that required to drive the engine is the total power developed in the cylinders. This is called the indicated horsepower (IHP). Brake horsepower, as we have seen, is measured by a mechanical device on the shaft; but since friction horsepower is involved, indicated horsepower cannot be measured directly. Fortunately, however, it can be calculated accurately, and its formula is:

$$IHP = \frac{PLANK}{33,000}$$

P stands for the Indicated Mean Effective Pressure (IMEP), the average pressure in the cylinder during the power stroke. This was originally measured by an instrument called an "indicator", hence the name. IMEP is expressed in pounds per square inch.

L is the length of the stroke in feet. P × A is measured in pounds, and L is in feet, so their product is foot-pounds of work done.

A is the area of the piston head in square inches. Since P is in pounds per square inch, and A is in square inches, the product of P and A is in pounds of force.

N is the number of power strokes per minute, per cylinder. For a four-stroke engine, N may be found by dividing the engine RPM by two, since only every other revolution has a power stroke.

K is the number of cylinders.

The product of P × L × A × N × K gives the number of foot-pounds of work done per minute by the engine.

To find the horsepower, divide the foot-pounds per minute by the constant, 33,000.

EXAMPLE:

A 6-cylinder aircraft engine has an indicated mean effective pressure of 125 PSI, a stroke of 5", and a bore of 5". What is the indicated horsepower at 2,750 RPM?

$$IHP = \frac{PLANK}{33,000}$$
$$= \frac{125 \times (5/12) \times (0.7854 \times 5 \times 5) \times (2750/2) \times 6}{33,000}$$
$$= 255.7 \text{ IHP}$$

4. Factors Affecting Engine Power

(a) Thermal Efficiency

The thermal efficiency of an engine is the ratio of the heat energy converted into useful work, to the heat energy contained in the fuel. Simply stated, it is the ratio of the power out, to the power in. If two engines produce the same amount of horsepower, but one burns less fuel than the other, the engine using the least fuel converts a greater portion of the available energy into useful work and therefore has the higher thermal efficiency.

Thermal efficiency may be found by this formula:

$$T.E. = \frac{Horsepower \times 33,000}{Pounds\ of\ fuel\ consumed\ in\ one\ minute \times 20,000 \times 778}$$

The horsepower used may be either brake or indicated, depending on the type of thermal efficiency you want. BHP gives brake thermal efficiency, and IHP, indicated thermal efficiency.

Thirty-three thousand is a constant, the number of foot-pounds of work per minute in one horsepower.

Fuel consumption is expressed in pounds of fuel burned in one minute.

Twenty thousand is the nominal heat energy content of aviation gasoline in Btu per pound.

Seven hundred and seventy-eight is another constant, the number of foot-pounds of work one Btu will produce.

EXAMPLE:

Find the brake thermal efficiency of an engine that produces 130 HP while burning 11.13 gal. of aviation gasoline per hour:

$$T.E. = \frac{130 \times 33,000}{(11.3 \times 6)/60 \times 20,000 \times 778} = 24.77\%$$

We can see from this, reciprocating engines are extremely inefficient with regard to the energy they

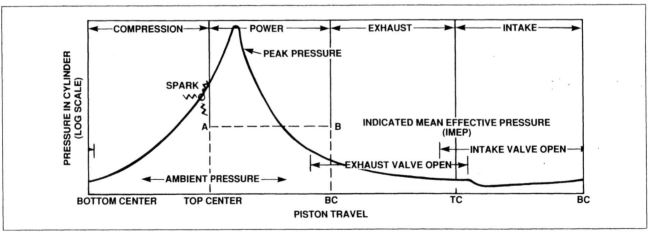

Figure 1-11. *The indicated mean effective pressure is the average pressure inside the cylinder during the power stroke.*

use. Somewhere between 25-30% utilization is normal.

b. Volumetric Efficiency

Reciprocating engines are air-breathing and require a maximum amount of air in the cylinder to release the most energy from the fuel. Volumetric efficiency is the ratio of the amount of air the engine takes into the cylinder to the total volume of the cylinder; and for this to be meaningful, the air in the cylinder must be of a standard density, which can be computed by correcting the temperature and pressure of the air to standard conditions, 59°F, or 15°C, and 29.92 in. Hg., or 14.69 PSI. If an engine draws in a volume of charge at this density exactly equal to the piston displacement, the volumetric efficiency of the engine will be 100%. A normally aspirated engine always has a volumetric efficiency of less than 100%, because the friction of the induction system walls and the bends in the tubing all restrict the amount of air that can flow into the cylinder during the time the intake valve is open. Supercharged engines, on the other hand, compress the air before it enters the cylinders, and they may have volumetric efficiencies greater than 100%. Volumetric efficiency may be found by the formula:

$$\text{V.E.} = \frac{\text{Volume of Charge}}{\text{Piston Displacement}}$$

Any of the following factors may lower the volumetric efficiency of a non-supercharged engine:

Part throttle operation: This restricts the amount of air that is allowed to flow into the cylinders.

Long intake pipes of small diameter: The friction increases directly as the length of the intake pipe and inversely as its cross sectional area.

Sharp bends in the induction system: The air will be slowed down each time it turns a corner, and the slower the air, the less will get into the cylinder.

Carburetor air temperature too high: As the temperature increases, the density of the air decreases, and there will be fewer pounds of air taken into the cylinder.

Cylinder head temperature too high: This also lowers the density of the air in the cylinder, providing fewer pounds of air for the same volume.

Incomplete scavenging: The incoming fuel-air mixture will be diluted with exhaust gases, and there will be less fresh charge drawn into the cylinder.

Improper valve timing: The intake valve will not be open long enough to draw the complete charge of fuel-air mixture into the cylinder.

c. Mechanical Efficiency

Mechanical efficiency is the ratio of the brake horsepower to the indicated horsepower; which shows the percentage of power developed in the cylinders that actually reaches the propeller shaft. Aircraft engines are usually quite efficient in this regard, and it is not unusual for 90% of the indicated horsepower to be converted into brake horsepower.

d. Piston Displacement

Piston displacement is the product of the area of the piston, the length of the stroke, and the number of cylinders; and since the amount of work done by the expanding gases is dependent on these factors, it is evident that the piston displacement is a major factor in power computations. Increasing either the bore of the cylinder, or the stroke of the piston will increase the piston displacement.

e. Compression Ratio

The ratio of the volume of the cylinder with the piston at bottom center to that with the piston at top center is called the compression ratio of the engine, and determines, to a great extent, the amount of heat energy in the fuel-air mixture that can be converted into useful work. By having as high a compression ratio as practical, the fuel-air mixture, when ignited, will release its energy rapidly, producing the maximum cylinder pressure at the point the crankshaft rotation will produce the most efficient push on the piston.

The practical limit to compression ratio is determined by the fuel used. When the compressed fuel-air mixture is ignited, the flame-front moves across the piston head, heating and further compressing the charge in front of it. A point is reached, called the critical pressure and temperature of the mixture, when it no longer burns, releasing its energy evenly, but rather explodes, releasing its remaining energy almost instantaneously. This is called detonation and creates such high pressures and temperatures that holes may be burned in piston heads, connecting rods kinked, or even cylinder heads blown off of the barrels. Fuels with high-octane ratings, a measure of critical pressure and temperature, allow high compression ratios to be used.

The pressure generated within the cylinder is important, as it must always be kept below the critical pressure of the fuel-air mixture. This cylinder pressure is determined by both the compression ratio and the pressure of the fuel-air mixture when it enters the cylinder. The pilot has no direct way of measuring the cylinder pressure, but they do have an instrument, the manifold pressure gage, that tells the pressure of the charge in the intake system. This pressure, usually measured in inches of mercury absolute, is related with the engine RPM to get an indication of the power being produced by the engine. For some power settings, there may be more than

1 RPM-manifold pressure combination, and the pilot, by trying each combination, can find the one that gives the most vibration-free operation for their particular aircraft.

f. Ignition Timing

It is of extreme importance that the maximum pressure within the cylinder be reached shortly after the piston passes its top center position and starts down. For this to occur, the mixture must be ignited quite a way before the piston reaches the top center, usually somewhere around 30° of crankshaft rotation before the piston reaches top center on the compression stroke.

Automobile engines have a variable timing device on their distributor that changes the amount of spark advance as engine operating conditions change; but aircraft engines employ fixed timing, a necessary compromise to give good performance for both takeoff and cruise power. When starting, the engine rotates so slowly that a spark occurring at the proper advanced time would cause a serious kickback and damage the starter. To prevent this, the spark is retarded for starting, using either an impulse coupling or a set of retard breaker points to time the spark from an induction vibrator.

If the ignition occurs too early, the engine will lose power because the maximum cylinder pressure is reached while the piston is still moving upward, and the force of the expanding gases opposes the rotational inertia of the engine. If, on the other hand, it occurs too late, there will be a more serious loss of power, since the cylinder volume is increasing as the gases expand and the effect of the push is lost. Also, late ignition does not allow enough time for all of the fuel-air mixture to be burned by the time the exhaust valve opens, and these burning gases, forced out past the exhaust valve, increase its temperature and will damage the engine by overheating.

The excess heat in the cylinder from late timing gives rise to a problem known as pre-ignition. The burning gases leaving the cylinder may cause local overheating of valve edges or carbon particles within the cylinder. When these particles glow from the heat, they will ignite the fuel-air mixture while it is being compressed, and cause, in effect, extremely early timing; so early, in fact, that the mixture will

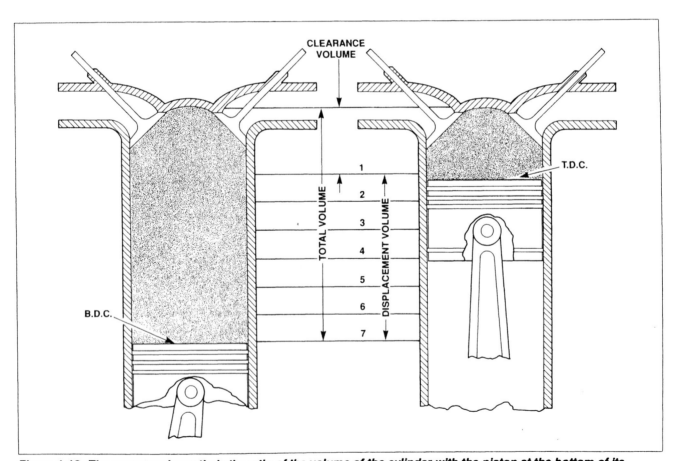

Figure 1-12. The compression ratio is the ratio of the volume of the cylinder with the piston at the bottom of its stroke to the volume with the piston at the top.

reach its critical pressure before the piston reaches top center, and detonation will occur.

Magnetos are timed to produce their spark at the proper position of the piston, but wear of the breaker points will cause the timing to drift early, or wear of the cam follower will cause the engine to rotate further before the points open, causing late timing.

g. Engine Speed

Power produced in an aircraft engine is determined by the cylinder pressure, the area of the piston, the distance the piston moves on each stroke, and the number of times this movement occurs in one minute.

Since the piston area, the length of stroke, and the number of cylinders are all fixed, the pilot has only two variables to work with; the pressure within the cylinder and the number of power strokes per minute. Engines equipped with fixed-pitch

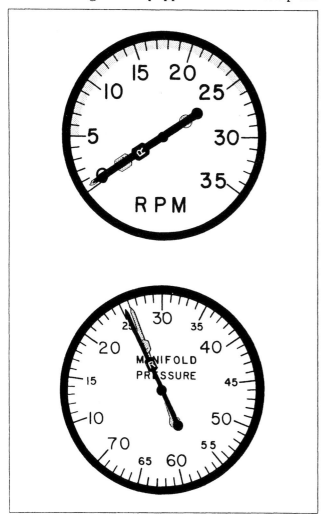

Figure 1-13. The tachometer and manifold pressure gage give the pilot the information needed to know the amount of power his engine is developing.

propellers give little control over the number of RPM the engine develops, and the maximum RPM for the rated power is determined by the airframe manufacturer by the choice of propeller. The propeller load limits the RPM, so the power setting is determined solely on the basis of RPM. If the engine is equipped with a constant-speed propeller, a governor will cause the propeller pitch to vary and maintain the RPM set by the pilot. The power is then determined by the relationship between the manifold pressure and the RPM.

Takeoff is made with the propeller in full low pitch, which allows the engine to reach its maximum RPM and develop its maximum power, and when the power is reduced, it is done by first reducing the amount of fuel by pulling back on the throttle, then reducing the RPM with the propeller pitch control.

If detonation should occur with any given power setting, the manifold pressure should be reduced and the RPM *increased*. This will distribute the energy released from the fuel over more power strokes, and the cylinder pressure will be reduced enough to stop the detonation.

h. Specific Fuel Consumption

While not actually a measure of the power itself, specific fuel consumption is an important measurement for comparing the efficiencies of engines. The number of pounds of fuel burned per hour to produce one horsepower, either indicated or brake, is known as the specific fuel consumption. For most practical purposes, brake horsepower is used and gives us brake specific fuel consumption.

Figure 1-14 illustrates the way specific fuel consumption varies with the RPM of the engine.

Since brake horsepower depends on engine speed below 2,200 RPM, the engine is not developing as much power as it is capable of for the amount of fuel it is using. Above 2,200 RPM, the amount of power required to drive the engine, the friction horsepower increases; and the brake horsepower decreases. From this, we see that the most efficient operating speed for this engine is around 2,200 RPM, at which speed the engine requires only about 0.48 lbs. of fuel per hour for every horsepower it produces. At full throttle, because of the additional power required to turn the engine itself, the power requirements have gone up to 0.55 lbs. per horsepower per hour.

5. Distribution of Power

When we consider the amount of power that is available in aviation gasoline compared to the amount of power actually delivered to the propeller shaft, we can easily see that an aircraft engine is an

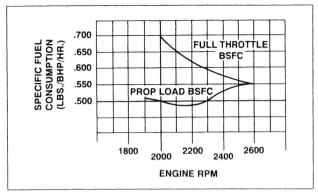

Figure 1-14. *The relationship between brake specific fuel consumption and engine RPM.*

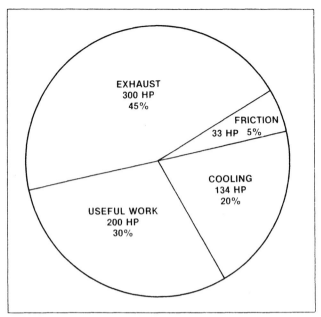

Figure 1-15. *The distribution of power developed by a reciprocating engine.*

inefficient machine. A typical engine may develop 200 BHP when burning 14 gal. of aviation gasoline per hour. But when this amount of aviation gasoline is burned in one hour, it releases enough heat energy to produce 667 HP. About 33 HP is used just to turn the engine and compress the air on the cylinders, and an equivalent of about 434 HP is lost in the air through the cooling and the exhaust systems. Anything that can be done to minimize the losses or increase the usable power must be done to increase the efficiency of the engine.

Even when power is delivered to the propeller shaft, we do not have what we really need for an airplane to be pulled through the air; we must convert the torque of the engine into thrust from the propeller. Torque, as we think of it with relation to a propeller, is a force which acts perpendicular to the axis of rotation of the propeller, and thrust acts parallel to its axis. A propeller converts only about 90% of the torque into thrust, and this is limited by the tip speed and by the inefficiency of the blade near the root.

Study Questions

1. What is a Btu?

2. What is a form of modern external combustion engine?

3. How does a rocket engine differ from a conventional internal combustion engine?

4. Why are two-stroke cycle engines not normally used for aircraft propulsion?

5. Where is the piston in a four-stroke cycle engine when each of these events occur?

    a. Intake valve opens

    b. Intake valve closes

    c. Exhaust valve opens

    d. Exhaust valve closes

    e. Ignition occurs

6. What is the difference between work and power?

7. How many watts of electrical energy is required to raise a 1,000 lb. weight, 6 ft., in 15 seconds?

8. What is the brake horsepower produced by a four-cylinder aircraft engine with the following dimensions and conditions? Bore 4", stroke 4" RPM 2,300, BMEP 125 PSI.

9. What is the brake thermal efficiency of an engine that burns six gallons of aviation gasoline per hour to produce 85 BHP?

10. What three factors affect the piston displacement of a reciprocating engine?

11. What determines the practical limit to the compression ratio of an aircraft engine?

12. How does late ignition cause overheating?

13. If detonation occurs in an engine equipped with a constant speed propeller, should the RPM be increased or decreased?

Chapter II

Construction And Operation

A. General Requirements

Everything that has been said to this point is applicable to most reciprocating engines. One must recognize, however, that there is a vast difference between the requirements of an aircraft engine and those of an automobile.

All aircraft engines must meet certain general requirements of efficiency, economy, and reliability; and often an engine becomes a compromise of requirements in order for the final product to meet a specific need.

1. Reliability

The ability of an engine to consistently do what the manufacturer says it will is the most important of all requirements of an aircraft engine. Regardless of what may be compromised to obtain the desired end results, reliability must never be sacrificed.

For an aircraft engine to be granted an Approved Type Certificate, it must prove its reliability by an endurance run witnessed by the Federal Aviation Administration in which it is operated for 85 hours at maximum continuous power, 15 hours at takeoff power, and 50 hours at high-cruise power. The engine is disassembled after its endurance run, and it must show no evidence of abnormal wear.

2. Durability

This is a measure of engine life obtained while maintaining the desired reliability. Time between overhauls of reciprocating engines in many cases is now more than 2,000 hours which, if equated to an automobile engine, would be about 250,000 miles before an overhaul is required. Time between overhauls, called "TBO" in the industry, varies in accordance with the percent of maximum continuous power used and by the prevailing operating conditions. For example, the engine in an airplane used for agricultural flying normally has a shorter TBO than an engine used for long flights at cruise power. On the opposite end of the operating schedule, engines that are operated for only an hour or two each week will need to be overhauled at a low time interval. This is because of the contaminants that form inside an engine when it is run, but not allowed to reach its operating temperatures throughout.

3. Compactness

Compactness is necessary to effect proper streamlining and balance of the aircraft, and in a single-engine airplane, the shape and size of the engine affects the visibility of the pilot.

4. Weight per Horsepower

This criterion has challenged engine manufacturers since Charles Taylor built the engine for the original Wright Flyer in 1903. His four-cylinder, water-cooled engine weighed about 180 lbs. and developed 15.76 HP as it warmed up. It had a weight-to-horsepower ratio of 15:1 in contrast with the modern Continental 0-200 engine which powers the Cessna 150 and weighs about the same, but develops 100 HP, giving it a weight-to-power ratio of 1.7:1.

5. Specific Power Output

It is the aim of the engine manufacturer to have his engine produce the greatest amount of power for the amount of fuel used. In order to increase the specific power output of an engine, the maximum amount of oxygen must be mixed with the fuel. This can be done by either compressing the air with a supercharger before it enters the cylinder, or by increasing the cylinder volume and compressing the fuel-air

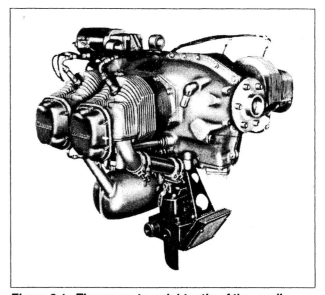

Figure 2-1. The power-to-weight ratio of the small modern aircraft engine is far better than that of its forerunners.

mixture more by using a higher compression ratio in the cylinders.

6. Fuel Economy

Fuel economy has improved as fuels have become more resistant to detonation, allowing the engine compression ratios to be increased. A modern engine with a compression ratio of 8.5:1, using aviation gasoline containing tetraethyl lead to suppress detonation, has a specific fuel consumption of 0.52 at cruise, which means that the engine burns 0.52 lbs. of fuel per hour for each horsepower produced. A 180 HP engine cruising with 65% power will burn 10.14 gals. of gasoline per hour.

7. Freedom from Vibration

Vibration may be minimized by using a large number of cylinders. This allows more overlap of the power impulses of the individual cylinders. This consideration, however, must be balanced by the efficiency of using fewer cylinders, and taking more power from each of them. To aid in damping vibrations, counter-weights and dynamic dampers may be installed on the crankshaft and special engine mounts used.

8. Operating Flexibility

This demands that an aircraft engine operate efficiently at both idle and at cruise, and it must never hesitate when full power is required for takeoff. It must operate at sea-level conditions and be capable of being adjusted to provide the required power at any altitude the airplane needs to fly. It must be designed so that the effects of rain, dust, sand, heat, and vibration are minimized.

9. Reasonable Cost

Reasonable cost is naturally a criterion of any piece of mechanical equipment. The first cost must be low enough for the engine to meet the competition of the market and be accepted by the airframe manufacturer, and the operating costs for its performance must be such that it will make a profit for its operators.

No single aircraft engine can be best in every area, and engines are built as a compromise in design from their inception.

B. Engine Configuration

1. Cylinder Arrangement

a. In-line

Figure 2-2 shows an engine popular in training aircraft as late as World War II.

The cylinders in this engine are arranged in a single line, and if the crankshaft is above them, the engine is said to be inverted. In-line engines were

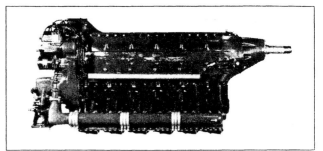

Figure 2-2. The in-line engine was popular for its small frontal area but its weight-to-horsepower ratio was not as favorable as that of the horizontally opposed engine.

popular in some of the early racing airplanes because of the small frontal area they presented. But since the crankshaft is the heaviest single component in an engine and this arrangement requires one throw for each cylinder, the shafts were long, and the weight-for-power, therefore, high. In-line engines may be cooled either by water or by air with air cooling the more popular. Air scoops brought the air into one side of the engine, and baffles directed it through the fins to remove heat from the cylinders. In-line engines were popular in four- and six-cylinder models with power outputs from 90 to about 200 HP.

b. V-type

A natural outgrowth of the in-line engine when more power was required was to place an additional cylinder on each throw of the crankshaft, and the V-type engine was the result. Two banks of cylinders are set apart by 45° or 60°.

One of the earliest aircraft engines to gain real notoriety was the Curtiss OX-5 of post World War I fame. This 90 HP, V-8, water-cooled engine was available in large quantities at low cost as military

Figure 2-3. The small frontal area of the V-engine made it popular in some of the high-speed fighter aircraft, but the complexity of its liquid cooling system caused it to pass from the scene.

surplus, and, although the engine was most inefficient, many airplanes were designed around it simply because of its availability.

The 450 HP Liberty V-12 engine served as a workhorse in the early 1920s — it, too, because of its availability, not its efficiency.

When World War II came along, airplane design had been developed to the point of requiring a powerful engine with small frontal area. Both Allison and Rolls Royce developed V-12 engines that produced in excess of 1,000 HP and powered some of the fastest fighter airplanes produced in both Britain and in the United States. Both the Allison and the Rolls Royce engines were liquid-cooled, and some models could be installed inverted. In one popular installation, the engine was installed behind the pilot and the propeller driven from a long extension shaft with a bevel-geared offset, allowing a cannon to be fired through the center of the propeller.

c. Radial

(1) Rotary-radial

The struggle to get the maximum power with as little weight as possible caused the designers of aircraft engines to concentrate on the crankshaft, the heaviest part of the engine. The logical step was to eliminate the shaft altogether, as was done with the rotary-radial engine, popular on both sides in World War I.

The propeller was mounted directly onto the crankcase along with the cylinders, while the pistons and rods were attached through slippers to an offset cam arrangement mounted on the airframe. The entire engine rotated, and as it did, the pistons moved in and out in the cylinders. The fuel-air mixture was admitted through a hollow tube in the center of the engine, and the burned gases were exhausted as the cylinder rotated beneath the fuselage.

Rotary engines turned rather slowly developing their full power at around 1,200 RPM, but even with this, the torque as well as other problems from all of the hardware thrashing around caused the designers to modify this engine into what became the static radial engine.

(2) Static Radial

A very small cast- or forged-aluminum alloy crankcase attaches to the airframe and holds a single-throw crankshaft. Mounted around this case are an odd number of air-cooled cylinders: three, five, seven, or nine. The piston in one cylinder attaches by a master rod to the crank throw, and all of the other pistons attach through link rods to the master rod. As the shaft turns, all of the pistons move up and down in their cylinders, and because all of these

Figure 2-4. The rotary-radial engine of World War I spun with the propeller while the pistons remained still. It has an excellent power-to-weight ratio but the mechanical problems led to its demise.

Figure 2-5. The static radial, or round engine, is still popular today for special applications where its favorable weight-to-power ratio is more important than its large frontal area.

engines use the four-stroke cycle, two revolutions are required to fire all of the cylinders. Having an odd number of cylinders gives an even number of degrees between each power application.

When more power was needed than could be produced in nine cylinders, the engines were doubled up, by using a crankshaft with two throws, and placing another bank of cylinders on the engine. 14- and 18-cylinder, 2-row radial engines have been very popular, and to further increase the number of cylinders using as small a frontal area as possible, four rows of cylinders were stacked onto one crankshaft. This made the 28-cylinder Pratt and Whitney R-4360, the largest practical radial engine used. The quest for more power from a reciprocating engine brought out one exceptionally large engine, a 36-cylinder, liquid-cooled, 4-row radial engine built by Avco-Lycoming; but this proved to be larger than was practical, and since the gas turbine engine came out at about the time this engine was designed, it was never put into production.

Figure 2-6. The Pratt and Whitney R-4360 was the largest radial engine to see practical service. Today, when this much power is needed, turbine engines are used.

Figure 2-7. The horizontally opposed cylinder configuration is the most popular arrangement today because of its light weight, its smooth operation and its small frontal area.

d. Horizontally Opposed

Through the process of evolution, the horizontally opposed cylinder arrangement has come out as the most popular configuration for the reciprocating engine. It has the advantage of much smaller frontal area than a radial engine of equivalent power, and by staggering the cylinders in the two rows, each connecting rod has its own crankshaft throw; yet the crankshaft can be much shorter than that of an in-line engine with the same number of cylinders.

Horizontally opposed engines have been built with 2, 4, 6, and 8 cylinders, with power outputs ranging from 37 HP to over 400. The smoothness of operation, small frontal area, and the high power for its weight have made the four- and the six-cylinder models the most popular engine for modern aircraft, both fixed and rotor-wing.

2. Cooling

Since it is possible to utilize only about 30% of the energy from burning aviation gasoline, the remaining 70% of the energy released must be disposed of. Some of this energy goes out the exhaust, and some of it is absorbed by the oil, but a large portion must be released into the air from the cylinders.

a. Liquid Cooling

An efficient way of removing unwanted heat from an engine is to surround the cylinders with a jacket containing a liquid into which the heat is transferred. This liquid is then pumped into a radiator from which the heat is transferred into the air. Liquid cooling keeps combustion chamber temperatures rather constant, which is much in its favor; but its use in aircraft brings with it many problems.

The weight of the radiators, plus all of the required pumps and plumbing and the coolant itself, adds to the weight of the powerplant. The complexity of the plumbing detracts from the engine reliability because of the possibility of leaks and pump failure, and the frontal area required by the radiators forces a drag penalty on the airplane. For these reasons, liquid-cooled engines have just about passed from the scene.

In World War II, the need for small frontal area in order to gain all of the speed possible, made the liquid-cooled V-12 engines popular for some of the high-speed fighter aircraft. These engines had a sealed cooling system and used ethylene glycol under pressure as a coolant. This allowed the engines to operate with a coolant temperature of around 110°C (230°F) and required very small radiators.

b. Air Cooling

Almost without exception, the engines used in current production general aviation aircraft are cooled by air. In the early days of air cooling, all that was necessary was for the finned cylinders to stick out into the airstream, and sufficient heat would be removed to keep their operating temperatures within acceptable limits. But now, the amount of power taken from each of the cylinders has been so increased that a more sophisticated cooling system such as pressure cooling has become a necessity. In pressure cooling, finned cylinders are encased in sheet metal baffles, and ram air pressure on one side of the engine forces air through the baffles and fins and removes a greater amount of heat than could be taken out by the fins alone.

The cylinder barrels of most air-cooled engines are made of chrome molybdenum steel with thin cooling fins machined all around their outside. The aluminum alloy head has its fins cast as an integral part, and the head is screwed and shrunk onto the barrel to form a gas-proof seal and afford the maximum amount of heat transfer. Areas of the head exposed to the most heat, such as around the exhaust valve, have additional fins cast onto them.

The engine compartment is divided by the cylinders with the upper half receiving ram air from the front of the cowling. Leather or plastic sealing strips prevent a loss of the ram air pressure between the compartment and the top cowling, and intercylinder baffles force the air through the fins where it picks up the heat.

The compartment below the engine has a low pressure caused by the deflection of the airstream, and some of the higher powered engines which require a closer control of the temperature have cowl flaps to control the air pressure below the engine. The lower this pressure, the more rapidly the air will be pulled through the fins, and the more heat will be removed from the cylinders.

3. Lubrication

Friction accounts for an appreciable percentage of the power losses in a reciprocating engine, and to minimize these losses, adequate lubrication systems are required. Almost all aircraft engines are lubricated with mineral oil which serves to minimize friction, transfer heat from the engine into the air, and form a seal between the fixed and moving parts of the engine.

a. Wet Sump Engines

The smaller engines with less stringent requirements on their lubrication system use a wet sump system in which all of the lubricating oil is carried in the engine itself in a compartment called the sump. A gear-type pump picks up the oil from the sump, and forces it through the bearings, into the valve operating mechanism, and throughout the engine as it is required; then it drains back down into the sump for the next cycle through the engine.

b. Dry Sump Engines

When a more elaborate lubrication system is required, or where more oil is needed than can be held in the engine, a dry sump system is used. The oil is carried in a tank outside of the engine and is fed into the oil pressure pump by gravity. The pump then forces the oil through the bearings and to all of the places where it is needed, and it then drains back into a small sump below the engine. A scavenger pump picks up the oil from the sump and returns it to the tank.

Since one of the functions of the oil is to remove heat from the inside of the engine, the oil picks up

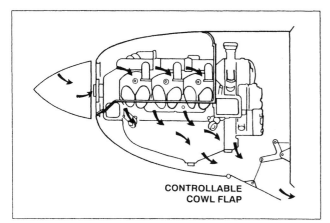

Figure 2-8. Pressure cooling is provided by ram air entering a compartment above the engine, flowing through the cylinder fins and exiting through cowl flaps at the lower portion of the engine cowling.

CONTROLLABLE COWL FLAP

Figure 2-9. A wet-sump engine carries its oil in a special compartment, or sump, which is an integral part of the engine.

the heat and carries it into an oil cooler where it is given up to the outside air.

C. Cylinder Numbering

1. Radial Engines

Number 1 cylinder on a radial engine is the top center cylinder, and the rest are numbered consecutively in the direction of crankshaft rotation. Twin-row radial engines are numbered in the same way, with all of the odd-numbered cylinders on the rear row and all of the even-numbered on the front row.

2. In-line and V-type Engines

Both in-line and V-type engines have their number 1 cylinder at the anti-propeller end, and the numbers increase toward the propeller. The left-hand bank of cylinders on a V-type engine are 1L through 6L, viewed from the anti-propeller end, and those on the right bank are 1R through 6R.

3. Horizontally Opposed Engines

There are two major manufacturers of opposed engines in the United States: Avco-Lycoming, and Teledyne-Continental. Engines of both manufacturers operate in the same way, but their cylinders are numbered differently. On Lycoming engines, number 1 is the front cylinder on the right side, and number 4 — or 6, depending on the number of cylinders in the engine — is the rear cylinder on the left side. Continental numbers their cylinders just the opposite. Number 1 is the rear cylinder on the right side, and 4 or 6 is the front cylinder on the left side. Both engines have all of the even-numbered

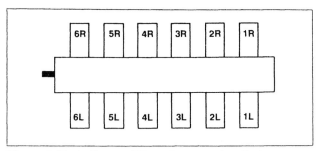

Figure 2-11. *In-line and V-engines number their cylinders with number one being at the anti-propeller end.*

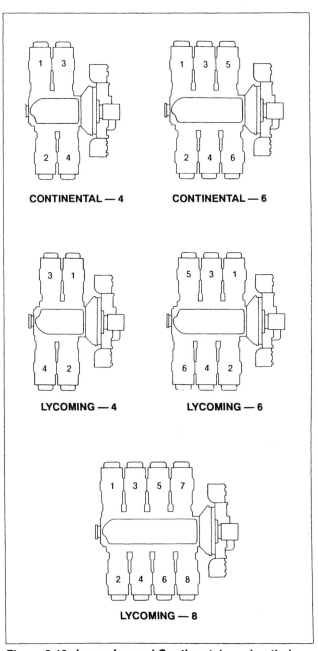

Figure 2-12. *Lycoming and Continental number their cylinders differently, but in either case cylinder number 1 is on the right side.*

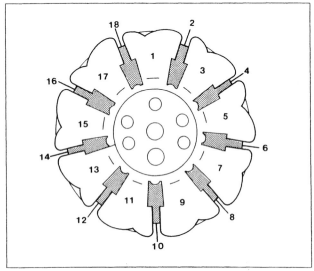

Figure 2-10. *The cylinders of a radial engine are numbered consecutively in the direction of rotation with number one being the top center cylinder.*

16

cylinders on the left side and all of the odd-numbered on the right.

D. Firing Order

1. Radial Engines

All 4-cycle engines must fire every one of their cylinders in two complete revolutions of the A crankshaft, and for the smoothest operation, the firing impulses must be evenly spaced. Radial engines having only one row of cylinders fire number 1 cylinder first, then each alternate cylinder in the direction of rotation. This makes the firing order of a 9-cylinder engine 1-3-5-7-9-2-4-6-8. The crankshaft rotates 80° between firing impulses.

Double-row radial engines fire alternate cylinders in each row. In an 18-cylinder engine, number 1 fires first, then number 12 which is across the engine in the front row. After this, the alternate cylinder in the rear row, or number 5, fires, and so on around the engine: 1-12-5-16-9-2-13-6-17-10-3-14-7-18-11-4-15-8. The

firing order for a 14-cylinder radial engine is 1-10-5-14-9-4-13-8-3-12-7-2-11-6.

A handy way to figure the firing order of an 18-cylinder radial engine is to begin with number 1; and to find the succeeding cylinders, add 11, then subtract 7. Continue doing this, either adding 11 or subtracting 7, to keep the numbers between 1 and 18. To find the firing order of a 14-cylinder, twin-row radial, add 9 and subtract 5.

2. In-line and V-type Engines

Six-cylinder in-line engines use a 120° crankshaft, such as the one seen in figure 2-14. Pistons 1 and 6 come to top center together, 1 on compression, and 6 on the exhaust stroke. After 120° of crankshaft rotation later, 2 and 5 come to top, and 120° later, 3 and 4. The firing order is 1-5-3-6-2-4.

V-type engines have the same configuration of crankshaft, with each throw having two rods attached to it by methods given in the description of connecting rods elsewhere in this text.

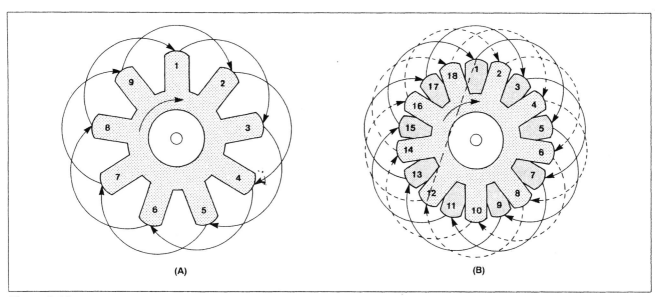

Figure 2-13.
(A) Firing order of a 9-cylinder radial engine
(B) Firing order of an 18-cylinder radial engine.

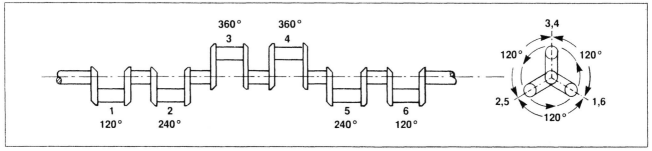

Figure 2-14. 6-cylinder, in-line engines use a 120° crankshaft.

3. Horizontally Opposed Engines

a. Four-cylinder

Four-cylinder, horizontally opposed engines, whether using the Lycoming or the Continental numbering system, fire in the same pattern, though their numbers are different. These engines use a 180° crankshaft; that is the throws are 180° apart, with 2 throws between each main bearing journal. The pistons in cylinders 1 and 2 are on top center at the same time, and 3 and 4 work together.

b. Six-cylinder

Six-cylinder, horizontally opposed engines use a 60° crankshaft, meaning that the six throws are arranged in three pairs, with each pair 60° apart. Pistons 1 and 2 come to top center together, then 3 and 4; and 60° later, 5 and 6. Since Continental and Lycoming put cylinder number 1 on the opposite ends of the engine, the firing impulses will be the same even though the firing order numbers are different. Lycoming engines fire 1-4-5-2-3-6, and Continentals fire 1-6-3-2-5-4.

c. Eight-cylinder

Eight-cylinder, horizontally opposed engines have four pairs of throws arranged with 90° between each pair. Pistons 1 and 2 are at top center at the same time 7 and 8 are on bottom. When 3 and 4 are on top, 5 and 6 are on bottom. All eight cylinders fire in 720° of crankshaft rotation, or 90° of crankshaft rotation between firing impulses. The firing order for the Lycoming 0-720 engine is 1-5-8-3-2-6-7-4.

E. Mechanical Systems

1. Cylinders

Power in an engine is developed in the cylinder, in the combustion chamber where the burning and expansion of the gases take place. The cylinder houses the piston and contains the valves through which the combustible mixture enters the cylinder and through which the exhaust gases leave.

A cylinder must be strong enough to withstand all of the internal pressures developed during engine operation, and to do this while at an elevated temperature. It must be lightweight in construction so that it will not pose a weight penalty on the engine; and while it must be designed and built so it will conduct the maximum amount of heat away from the engine, it must be relatively simple to build, inspect, and maintain.

Liquid-cooled cylinders are usually arranged in banks, or blocks, to facilitate the flow of the coolant; but for ease of maintenance and the ability to replace one cylinder if it becomes damaged, individual cylinders are almost universally used for air-cooled engines.

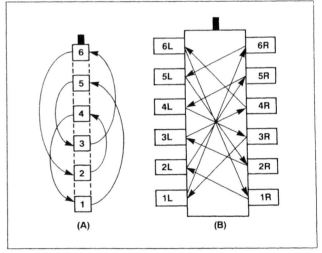

Figure 2-16.
(A) Firing order of a 6-cylinder in-line engine.
(B) Firing order of a V-12 engine.

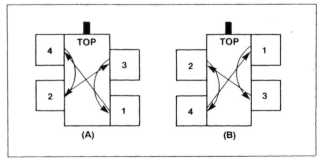

Figure 2-17.
(A) Firing order of a 4-cylinder Continental, horizontally opposed engine.
(B) Firing order of a 4-cylinder Lycoming, horizontally opposed engine.

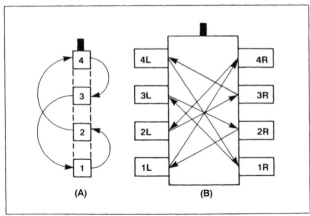

Figure 2-15.
(A) Firing order of a 4-cylinder, in-line engine.
(B) Firing order of a V-eight engine.

18

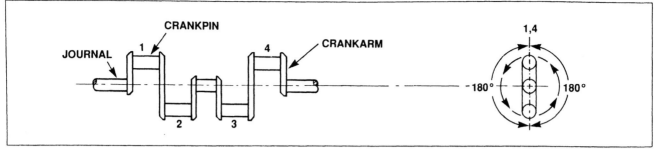

Figure 2-18. 180° crankshaft used on a 4-cylinder, horizontally opposed engine.

a. Cylinder Barrels

A high-strength chrome molybdenum steel barrel is machined with a skirt which projects into the crankcase, a mounting flange to attach the cylinder to the case, thin cooling fins, and, on most cylinders, threads to screw into the cylinder head.

Rather than having the cylinder bore ground perfectly straight, most cylinders have a certain degree of choke. This means that at the point where the heat is most concentrated, in the area of the head, the bore diameter is slightly less than that in the main portion of the cylinder. By the time the cylinder reaches operating temperature, the choked area has expanded more than the lower part of the barrel, and the bore has become straight. If the cylinder were not ground with this choke, at the operating temperatures, the top would be larger than the bottom, and there would be danger of the combustion gases leaking past the piston rings and damaging the piston.

The piston and its rings spend their useful life rubbing up and down on the cylinder wall, wearing it away. To minimize this wear, some cylinder walls

are hardened. There are two commonly used procedures for providing a hard surface: one is chrome plating, and the other, a form of case hardening known as nitriding.

(1) Chrome-plated Barrels

Chromium is a hard, natural element which has a high melting point, high heat conductivity, and a very low coefficient of friction, about one half that of steel. The cylinder barrel is prepared for chrome plating by grinding to the required size and submerging it in a plating solution where a coating of chromium is electrolytically deposited on the inside of the barrel. This chromium has a natural tendency

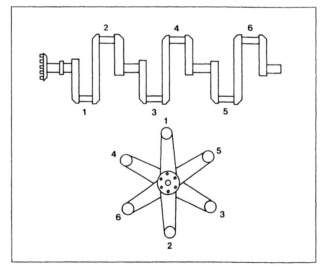

Figure 2-20. 60° crankshaft used in a 6-cylinder, horizontally opposed engine.

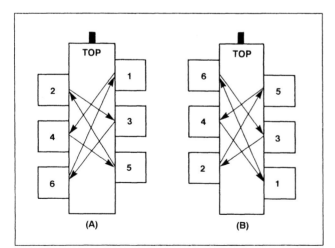

Figure 2-19.
   (A) Firing order of a Lycoming 6-cylinder, horizontally opposed engine.
   (B) Firing order of a Continental 6-cylinder, horizontally opposed engine.

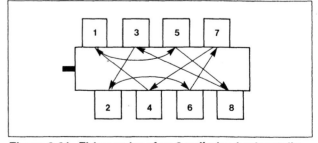

Figure 2-21. Firing order of an 8-cylinder, horizontally opposed engine.

to form surface cracks, and by a carefully controlled process, these almost microscopic cracks are opened up until they form a network of visible and interconnected cracks or channels that hold lubricating oil on the cylinder wall.

There are many advantages of chromed cylinders over either plain or nitrided steel. They are less susceptible to rusting or corrosion, both because of the natural corrosion resistance of chromium, and because of the tendency of the oil to adhere to the chromed cylinder walls better than to plain steel. The hardness of chromium causes less wear of the wall than would occur on a softer cylinder, and when the wall does wear, the cylinder may be ground enough to straighten it, and then it may be re-chromed back to standard size. Cylinders with chrome-plated walls are identified by an orange band around their base or by a stripe of orange paint on certain of their fins.

(2) Nitrided Barrels

Case hardening is a process in which the surface of steel is changed by the infusion of some hardening agent. It differs from plating in that there is no material on the surface, but actually a change in the surface material itself. Nitriding does not require quenching, and it does not warp the cylinder as other forms of case hardening might do. After the cylinder barrel has been ground to the required size and smoothness, it is placed in a furnace, or retort, in an atmosphere of ammonia gas. The length of treatment and the temperature are both carefully controlled, and in the process, the ammonia gas (NH_3) breaks down, or disassociates, into nitrogen and hydrogen. The steel used in the cylinder barrel has a small percentage of aluminum as an alloying agent, and the nitrogen combines with this aluminum to form aluminum nitrides, the hard, wear-resistant surface we want. Since nitriding is not a plating or a coating, it causes a dimensional growth of only about 0.0002" to 0.0004", but the hardened layer varies in depth to about 0.002", and gradually decreases in hardness from the surface inward, until it corresponds with that of the metal itself. After the nitriding process is completed, the cylinder walls are honed to a microsmooth finish.

One of the problems with a nitrided surface is its susceptibility to corrosion or rust. Nitrided cylinder walls must be kept covered with oil, and if an engine is left out of service for any period of time, the walls should be coated with a sticky preservative oil.

Nitrided cylinder walls are identified by a band of blue paint around their base, or by certain fins being painted blue.

b. Cylinder Heads

Most air-cooled aircraft cylinder heads are sand-cast of an aluminum alloy containing copper, nickel, and magnesium. This alloy has the desirable characteristics of relatively high strength and the ability to

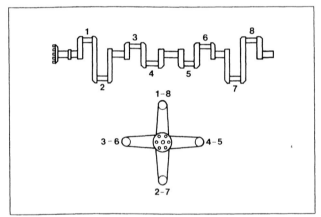

Figure 2-22. 90° crankshaft used in an 8-cylinder, horizontally opposed engine.

ENGINE TYPE	FIRING ORDER
4 CYLINDER INLINE	1-2-4-3 or 1-3-4-2
6 CYLINDER INLINE	1-5-3-6-2-4
12 CYLINDER V	1L-6R-5L-2R-3L-4R-6L-1R-2L-5R-4L-3R
4 CYLINDER OPPOSED (LYCOMING)	1-3-2-4
4 CYLINDER OPPOSED (CONTINENTAL)	4 1-4-2-3
6 CYLINDER OPPOSED (LYCOMING)	1-4-5-2-3-6
6 CYLINDER OPPOSED (CONTINENTAL)	1-6-3-2-5-4-
14 CYLINDER RADIAL	1-10-5-14-9-4-13-8-3-12-7-2-11-6
18 CYLINDER RADIAL	1-12-5-16-9-2-13-6-17-10-3-14-7-18-11-4-15-8
28 CYLINDER RADIAL	A1-B5-C2-D6-A3-B7-C4-D1-A5-B2-C6-D3-A7-B4-C1-D5-A2-B6-C3-D7-A4-B1-C5-D2-A6-B3-C7-D4

Figure 2-23. Firing orders of various configurations of aircraft engines.

20

maintain this strength to temperatures of up to about 600°F.

Cooling fins are cast into the surface of the head to increase its cooling area, and because of the difference in temperatures at the various sections of the cylinder head, it is necessary to provide more cooling area in some sections than others. The exhaust valve region is the hottest part of the head, and there are more fins there than anywhere else on the head.

The holes for the spark plugs are bushed with either bronze or steel bushings screwed, shrunk, and pinned in place, or by stainless steel Heli-coil® inserts. Bronze, cast iron, or steel valve guides are shrunk in place, and hardened, ring-type valve seats are shrunk in the head to provide a wearing surface for the intake or exhaust valves and to protect the aluminum alloy of the head from the hot exhaust gases.

Most cylinder heads are heated to expand them, and then screwed onto the steel cylinder barrels which have been chilled to contract them. When the two reach the same temperature, they fit so tightly that there is no leakage of the hot gases. The Teledyne-Continental Tiara engine, rather than screwing their cylinder heads onto their barrels, uses long through-bolts to hold the head in place as well as to hold the cylinder to the engine.

c. Valve Mechanism

(1) Valves

The valves in the cylinders of an aircraft engine are subject to high temperature, corrosion, and operating

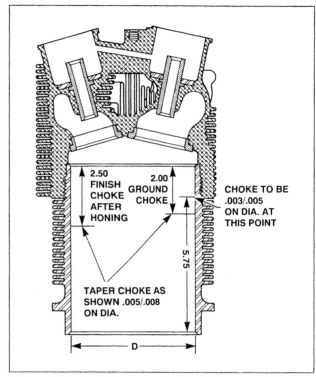

Figure 2-25. A choke-ground cylinder has a smaller diameter at the head end than at the skirt to allow for the uneven expansion caused by the mass of the cylinder head. The bore becomes straight at operating temperature.

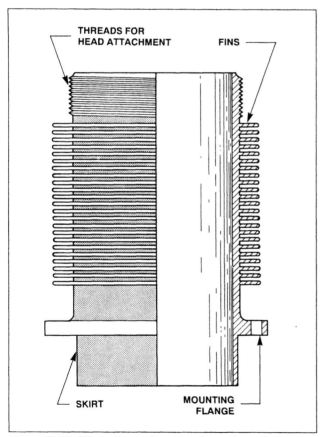

Figure 2-24. The cylinder barrels of an air-cooled aircraft engine are machined from high-strength chrome molybdenum steel.

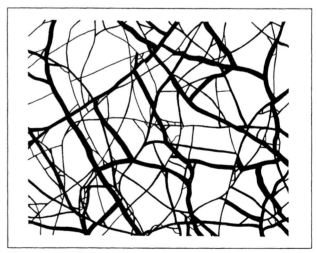

Figure 2-26. The walls of a chrome plated cylinder are covered with thousands of tiny cracks that hold the lubricating oil.

stresses; and the metal used in their manufacture must be able to resist all of these attritional factors.

Because intake valves operate at lower temperatures than exhaust valves, they may be made of chrome nickel, or tungsten steel, while the exhaust valves are usually made of some of the more exotic alloys such as inconel, silicon-chromium, or cobalt-chromium alloys.

The ground face of the valve forms a seal against the valve seat in the cylinder head when the valve is closed.

The face is usually ground to an angle of 30°, 45°, or 60° with the choice made that will give the best airflow efficiency and seating.

Valve faces are often made more durable by the application of a material called Stellite. About 1/16″ of this alloy is welded to the valve face and ground to the correct angle. Stellite is resistant to high temperature and corrosion, and also withstands the shock and wear associated with valve operation.

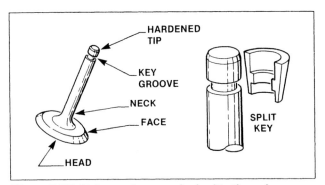

Figure 2-27. Valve springs are locked to the valves with tapered split keys.

The surface-hardened valve stem acts as a pilot for the valve head and rides in the valve guide installed in the cylinder head for this purpose. The tip of the stem is hardened to withstand the hammering of the valve rocker arm as it opens the valve, and a groove machined around the stem near the tip holds the split-ring stem keys which hold the valve spring retaining washer in place.

Some exhaust valve stems are hollow and partially filled with metallic sodium. The sodium melts at approximately 208°F, and the up and down motion of the valve circulates the liquid sodium so it can carry heat from the valve head into the stem, where it can be dissipated through the valve guide to the cylinder head and then into the cooling fins. In this way, the operating temperature of the valve can be reduced by as much as 300-400°F.

Most intake valves have either flat or tulip-shaped solid heads, and in some engines, the intake and exhaust valves are similar in appearance; they are not interchangeable, however, as they are constructed of different materials.

(2) Valve Seats

The valve operates in the hot environment of the inside of the cylinder head and is exposed to continual pounding; and for this reason an extremely durable valve seat must be installed in the soft aluminum cylinder head. Rings of aluminum-bronze or steel are machined with an outside diameter

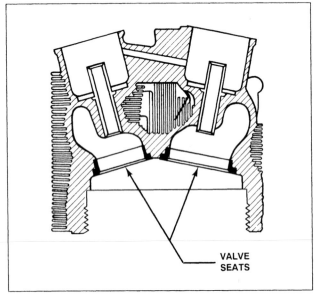

Figure 2-28. Valve seats made of aluminum bronze or steel are shrunk into place in the soft aluminum head.

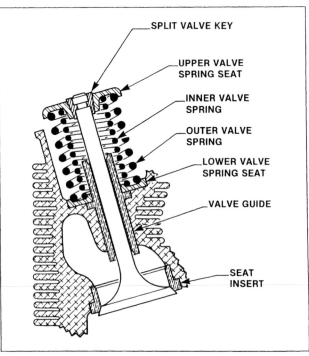

Figure 2-29. Bronze or cast iron valve guides are shrunk into the cylinder head to guide the valves.

about 0.010" to 0.015" larger than the hole into which they are to fit, and the head is heated in an oven to around 575-600°F. and the seat put in place. The interference fit will then hold it in place when the head and seat are at the same temperature.

The seats are ground with a high-speed stone guided by a pilot through the valve guide so the seat will be absolutely concentric with the valve stem.

(3) Valve Guides

The hardened valve stem rides in the cylinder head in bronze or cast-iron guides that are chilled and pressed into the heated cylinder head. The stem receives the heat from the valve head and transfers it into the guide, where it is taken into the cylinder head and then dissipated into the air. Because of this heat transfer, it is important that the tolerance allowed by the manufacturer for the stem-to-guide clearance is not exceeded. This fit is provided by reaming the guide after it has been pressed into place.

After the valve seat and guide have been properly installed in the cylinder head, a pilot is slipped through the guide and the seat ground soft is concentric with the valve guide hole.

(4) Valve Springs And Retainers

The valves are opened by the rocker arm pushing on the valve stem tip, but they are closed by two, or sometimes three, coil springs. Multiple springs are used, not only because of the additional force they provide, but because of a characteristic of the spring known as resonance. Valve opening is done with a series of impulses, and at a certain engine RPM, these impulses will occur at the resonant frequency of the spring. When this happens, the spring loses its effectiveness and allows the valve to float. To prevent this floating condition, two or more springs having different pitch, diameter, and wire size are used; and because of their different configuration, they have different resonant frequencies, so the engine can operate throughout its full range of RPM without valve float problems.

The valve spring seat is installed around the valve stem, and the two springs are slipped in place. An upper valve spring seat is put over the springs and they are compressed. Two split valve keys are installed in the groove in the valve stem. When the springs are released, the upper springs seat will force, and hold, the valve keys in the stem groove and hold the valve tightly closed.

2. Pistons

The piston in a reciprocating engine is a cylindrical device which moves back and forth within the cylinder, acting as a moving wall for the combustion chamber.

As the piston moves inward in the cylinder it creates a low pressure and draws in the fuel and air mixture. As it moves outward, it compresses the charge, and when ignition occurs, the expanding gases force the piston back inward. This force is transmitted to the crankshaft through the wrist pin and connecting rod, and on the next outward stroke, the piston pushes the burned gases from the cylinder.

The majority of aircraft pistons are machined from aluminum alloy forgings, and they have grooves cut into their outer surface to receive the piston rings. The underside of the head is usually finned to aid in cooling and to provide extra strength.

Almost all aircraft pistons are of the trunk-type, like those seen in figure 2-30, with the top, or the head, either domed or flat; many have recesses machined into their heads to provide more clearance for the valves.

The piston pin boss is an enlarged portion of the skirt which is machined to fit the piston pin, with the additional material providing the strength needed to transmit the force of the expanding gases to the connecting rod.

The compression ratio of the engine may be varied by changing pistons, and for this reason it is vital that the proper part number be used any time a piston is replaced in an engine. Since the piston is the reciprocating part of an engine, the inertia involved in its continual starting and stopping will cause engine vibration if there is any difference in piston weights. Because of this, the manufacturing

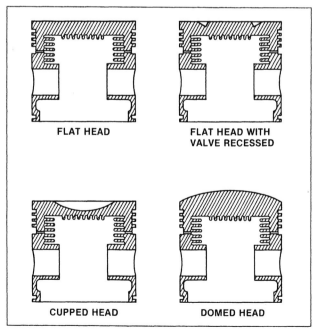

FLAT HEAD

FLAT HEAD WITH VALVE RECESSED

CUPPED HEAD

DOMED HEAD

Figure 2-30. Head shapes for aircraft engine pistons.

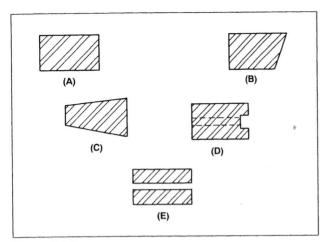

Figure 2-31.
- (A) Rectangular piston ring.
- (B) Tapered piston ring.
- (C) Wedge-type piston ring.
- (D) Single-piece oil control ring.
- (E) Two-piece oil control ring.

tolerance for aircraft pistons requires that they be held to within about a quarter of an ounce (seven grams) of each other.

a. Piston Rings

The piston serves as a plunger to compress the fuel-air charge and transmit the heat from the burning gases into the cylinder wall, but the wide variation of piston temperatures causes such dimensional changes that it is impossible to have a solid plug fit in the cylinder. Grooves are machined around the piston and fitted with cast-iron rings, expanded so they will fit against the bore of the cylinder with considerable pressure. The top two or three grooves are fitted with solid rings called compression rings, whose primary purpose is to provide a seal and

prevent gases escaping around them, and to transfer heat into the cylinder wall.

Oil control rings are installed immediately below the compression rings. They are often of the two or more piece construction and may have corrugated expanders behind them in the ring groove. These grooves often have holes drilled in them to drain some of the oil back into the crankcase. The purpose of the oil control ring is to maintain the proper quantity of oil between the piston and the cylinder wall.

At the very bottom of the piston skirt, oil scraper or wiper rings are often installed. These are usually tapered or beveled rings and may be installed in such a way that they wipe the oil either toward or away from the piston head, depending on the engine. The installation of these rings is critical and must follow the manufacturer's instructions in detail.

Piston rings are usually made of gray cast iron and may have one of several cross-sectional shapes such as those seen in figure 2-31.

Wedge-shaped piston rings are more or less self-cleaning and reduce the carbon accumulation in the grooves.

Newly installed piston rings do not form a good seal with the cylinder wall, and they must be seated, or worn in, before they will stop high oil consumption. The high pressure and high temperature found in the combustion chamber of an aircraft engine make it extremely important that only the piston ring approved by the manufacturer be used. The end-gap clearance must be within the tolerance specified, the tension proper, and the side clearance between the ring and the groove as recommended.

Piston rings used in cylinders with plain steel walls may be chrome-plated for better wear characteristics,

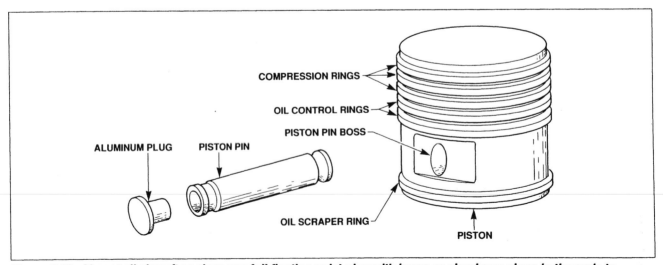

Figure 2-32. Almost all aircraft engines use full floating wrist pins with brass or aluminum plugs in the ends to prevent their scoring the cylinder walls.

but rings used on chrome-plated or nitrided cylinder walls must be of plain cast iron, since hardened piston ring faces will not seat in a hardened cylinder barrel, and the engine will not break in and stop pumping oil.

b. Piston Pins

Hollow, hardened alloy-steel pins fit into the hole in the piston pin boss, and attach the piston to the connecting rod. Most of the piston pins — or wrist pins as they are called in modern aircraft engines — are of the full-floating type; that is, they are free to rotate in both the piston and the connecting rod, and soft aluminum or brass plugs fit into the ends of the pin to prevent it scratching the cylinder wall as the piston moves in and out.

Piston pins are usually a push fit into the piston and are lubricated by oil flowing through holes drilled in the bosses of the piston.

3. Connecting Rods

The connecting rod is the link which transmits the forces between the piston and the crankshaft. It must be strong enough to remain rigid under load, and yet be light enough to reduce the inertia forces which are produced when the rod and piston stop,

change direction, and start again at the end of each stroke.

In order to attach all of the pistons in one row of a radial engine to one throw of a crankshaft, a master rod with link or articulating rods is used. In figure 2-33(A), we see a typical 9-cylinder arrangement. The master rod attaches to the single throw of the crankshaft, and the eight link rods connect their respective pistons to the master rod through knuckle pins.

V-type engines often have two cylinders directly opposite each other, and both cylinders must be attached to the same throw of the crankshaft. One of the more popular ways of doing this is by using the fork-and-blade-type rod. One piston is attached to the rod having its forked big end around the crank pin with a 2-piece bearing insert installed in the rod. The blade-type rod attaches to the opposite piston, and its big end rides on the outside of the bearing insert in the fork.

Horizontally opposed engines, rather than sharing a crank pin between two cylinders, stagger the cylinders so each one can have its own crank pin. The end of the rod attached to the crank pin is fitted with a cap and a 2-piece bearing. The bearing cap is

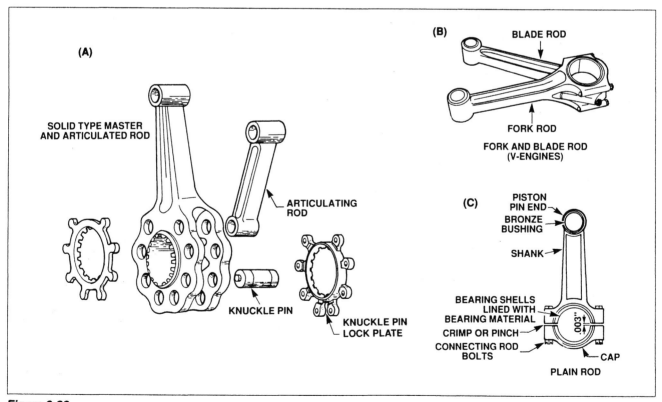

Figure 2-33.
- (A) Master and link rods for a radial engine.
- (B) Fork-and-blade connecting rods for a V-type engine.
- (C) Two-piece connecting rod for horizontally opposed engine.

held on the end of the rod with bolts and locknuts. The cap and rod assembly are paired, and must always be installed as a matched set.

4. Crankshaft

The crankshaft is the backbone of a reciprocating engine and is subjected to most of the forces developed by the engine. Its main purpose is to transform the reciprocating motion of the piston and connecting rod into rotary motion for turning the propeller. The crankshaft, as the name implies, is a shaft composed of one or more cranks, or throws, located at specified points along its length. These throws are formed by forging off-sets into a shaft before it is machined. Since crankshafts must withstand high stresses, they are generally forged from a very strong alloy steel, such as chrome-nickel-molybdenum.

a. Design

Since the crankshaft is the heaviest single part of an aircraft engine, the engine's evolution has been largely dictated by the crankshaft.

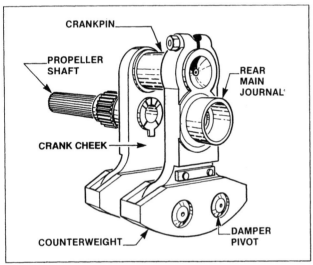

Figure 2-34. Two piece crankshaft for single-row radial engine.

In-line engines must have long crankshafts with each cylinder having its own throw, often with a main bearing between each cylinder. This makes them heavy. V-type engines partially solved this weight problem, as two cylinders share a single crankshaft throw. But the ultimate in weight savings is the radial engine, with only one throw for each row of cylinders. Figure 2-34 shows a typical single-row radial engine crankshaft. Single-row radial engines may have three, five, seven, or nine cylinders, and when it is needed to get more power than can be obtained from a single row, two or even four banks of cylinders may be stacked and two- or four-throw crankshafts used.

By far, the most popular configuration of engine today is the horizontally opposed engine with a crankshaft having a throw for each cylinder. The throws are paired and the cylinders slightly staggered. Figure 2-35 shows a crankshaft of a 4-cylinder opposed engine. This crankshaft has four throws and three main bearings. The two pairs of throws are 180° apart, so that when the piston of number 1 cylinder is at the top of its stroke, number 4 will be at the bottom; and when 2 reaches the top, 3 will be on the bottom.

Six-cylinder opposed engines use what is called a 60° crankshaft, meaning that the throws are 60° apart. In figure 2-36, we have a typical 6-cylinder crankshaft with four main bearings and three pairs of throws. The throws for cylinders 1 and 2 are opposite each other, and 3 and 4 are opposite, and are separated by 60° from the 1-2 pair. The pair of throws for cylinders 5 and 6 are 60° from those of 3 and 4.

b. Construction

Aircraft crankshafts are forged of high-strength alloy steel, drilled for oil passage, and the main bearing journals and the crank pins are ground and polished and then the entire crankshaft nitrided. The hollow crank pins have thin steel, spool-like plugs pressed

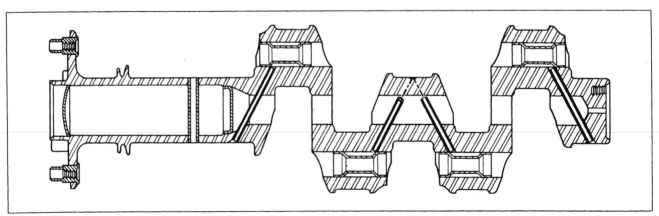

Figure 2-35. Crankshaft for a 4-cylinder, horizontally opposed engine.

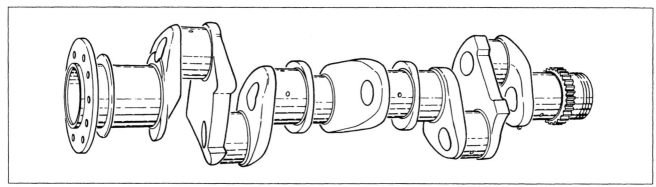

Figure 2-36. Crankshaft for a 6-cylinder, horizontally opposed engine.

in place. These sludge plugs, as they are called, allow the centrifugal force of the spinning crankshaft to trap any sludge that is in the oil and prevent its circulation through the engine. The sludge is removed when the engine is overhauled and the plugs are replaced.

c. Dynamic Dampers

Crankshafts are made as light as possible in keeping with their strength requirements, and they are balanced both statically and dynamically. The force that causes the crankshaft to rotate is not a smooth push, as would be desired, but is rather a series of pulses or intermittent pushes. At certain rotational speeds of the engine, the frequency of these pulses can be the same as the natural resonant frequency of the crankshaft and propeller combination, and the forces generated could be strong enough to cause damage to the engine. To prevent this, dynamic dampers are installed to change the natural resonance of the crankshaft-propeller combination to some value outside that which can be excited by the power impulses.

In figure 2-37, we see the way dynamic dampers operate. We know that a pendulum is a body suspended from a point and is free to swing back and forth. It has a natural period — that is, one frequency that will require the minimum amount of excitation to cause it to swing. This period is determined by both the length and the mass of the pendulum. Now if the pendulum rod were hinged in its center and another weight installed at the hinge, the energy pulses would not cause the same swings they had previously because the length has been changed as well as the mass.

Aircraft crankshafts have counterweight blades forged as an integral part of the shaft, and counterweights fit over these blades. The pins that hold the counterweights to the blades are considerably smaller than the holes in the blades, so they form a

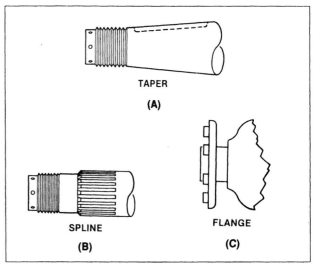

Figure 2-38.
(A) Low powered engines may use a tapered propeller shaft to hold the propeller hub.
(B) Splined shafts have been used in many aircraft engines to hold the propeller.
(C) Most modern aircraft engines have a flange forged as an integral part of the propeller shaft to which the propeller is bolted.

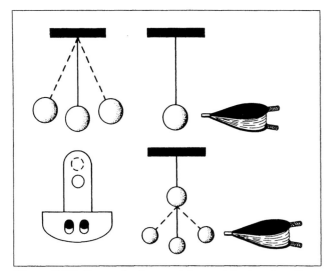

Figure 2-37. Dynamic dampers are used on many of the more powerful engines to change the resonant frequency of the rotating mass and minimize torsional vibrations.

loose fit and the weight is free to oscillate back and forth. When the engine is running, centrifugal force holds the counterweights out and the vibrations from the power impulses are absorbed by the counterweights rocking back and forth in the holes.

d. Propeller Attachment

As with many parts of an aircraft engine, the method of attaching the propeller to the crankshaft has undergone an evolutionary process. The earliest engines used a tapered propeller shaft with a single steel key, figure 2-38(A), and the wooden propellers were mounted in a steel hub that fitted over the taper. The front end of the shaft was threaded to accept the propeller retaining nut which was a part of the hub. A snap ring inside the hub acted as a puller when the retaining nut was backed off, and this made it possible to remove the propeller from the shaft.

As engine power increased, the need was felt for a stronger attachment of the propeller to the crankshaft, and a splined shaft was adopted by almost all of the engine manufacturers. A split, single-piece, bronze rear cone centers the rear of the propeller on the shaft, and a 2-piece steel front cone holds the front of the hub centered. The retaining nut fits into a groove in the front cone, and a snap ring inside the propeller hub allows the nut and the front cone to act as a puller for the propeller.

Most modern engines use a flanged crankshaft to mount the propeller. Bolts and nuts secure the propeller to the flange which is forged as an integral part of the crankshaft. Controllable pitch propellers require oil for their operation, and this oil is fed into the propeller through the hollow crankshaft. O-rings seal the crankshaft to the propeller hub and prevent leakage of propeller control oil.

5. Propeller Reduction Gearing

Earlier in this book, we saw that the power developed by an aircraft engine is dependent on the force applied by the expanding gases on the piston and the distance the piston is moved in a given period of time. As the power impulses per minute, the RPM, are increased, the power output increases. But, when the engine is speeded up, we run into other problems such as the tip speed of the propeller approaching the speed of sound. As it reaches this speed, its efficiency drops off radically.

High engine RPM and low propeller speed may be attained by using a reduction gearing arrangement on the engine. This naturally adds weight and complexity, and robs some of the power, but the overall gain is larger than the sum of the losses, and so geared engines are commonly used when power requirements are high.

The most simple type of reduction gearing is the spur gear, figure 2-39(A), attached to the crankshaft.

This drive gear meshes with a similar, but larger, gear to which the propeller is attached. This arrangement has a few drawbacks, but has been successfully used on such engines as the Continental GO-300. The crankcase must be made quite strong to withstand the precessive forces put into it by the gyroscopic action of the propeller, and the engine must rotate in the direction opposite the propeller. The propeller shaft is mounted quite high, and unless the installation is carefully designed, it can reduce the forward visibility of the pilot.

One method of overcoming some of the problems of the simple spur gear has been the use of an internal-tooth-driven gear, with an external-tooth-drive-gear on the crankshaft, figure 2-39(B).

Torsional vibration problems are compounded when an engine is geared, and in addition to the counterweights on the crankshaft, some engines use

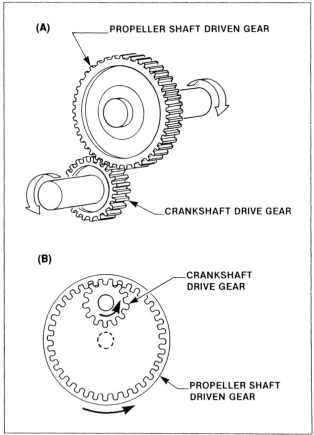

Figure 2-39.
- **(A)** *The spur-type reduction gear arrangement is the simplest method of reducing the RPM of the propeller relative to crankshaft speed.*
- **(B)** *If the spur-type drive gear drives an internal tooth driven gear, the direction of the propeller is not reversed.*

a quill shaft to further reduce these vibrations. One end of the quill shaft, figure 2-40, is splined into the front end of the crankshaft, and its opposite end is splined into the front end of the propeller drive gear. In this way, instead of the propeller drive gear attaching solidly to the crankshaft, it attaches through the quill shaft which flexes enough to absorb some of the shocks.

The new series of engines produced by Teledyne-Continental, known as the Tiara engines, use a unique application of the quill shaft to minimize crankshaft vibrations. These vibrations change as the engine RPM changes, and to provide the proper stiffness of the reduction gearing as the RPM changes, a Vibratory Torque Control is used. At low engine speeds, the propeller shaft drive gear is rigidly attached to the crankshaft; but as the engine speed increases to the range where a flexible drive would reduce the vibrations, the VTC unit automatically disengages the drive gear from the crankshaft and drives it through a quill shaft.

High output engines use planetary gears. With this type of arrangement, the propeller shaft turns in the same direction as the engine crankshaft, and the propeller shaft is in line with the crankshaft.

There are two basic types of planetary gear arrangements in common use. In figure 2-41(A), we see one type, in which the ring gear is fixed into the nose section of the engine and the crankshaft turns the sun gear. The planetary gears, or pinions, spin on stub shafts mounted on the rear end of the propeller shaft. The speed reduction of this type of gear arrangement may be figured by the formula:

$$\text{Gear ratio} = \frac{\text{teeth on sun gear} + \text{teeth on ring gear}}{\text{teeth on sun gear}}$$

Another type of planetary gear system uses bevel gears, and in this arrangement, the sun gear is attached to the nose section of the engine and the ring is driven by the crankshaft. The propeller shaft mounts on a spider that holds the planetary gears and the propeller turns more slowly than the engine by the ratio found by the formula:

$$\text{Gear ratio} = \frac{\text{teeth on ring gear} + \text{teeth on sun gear}}{\text{teeth on ring gear}}$$

The direction of rotation of the propeller is the same as that of the crankshaft.

Normally the power developed by an aircraft engine is determined by the pilot using the appropriate settings of the manifold pressure and engine RPM, but certain large radial engines had what was known as a torque nose. The ring gear was equipped with a hydraulic cylinder that generated a pressure proportional to the torque produced by the engine, and this pressure was read by the flight engineer on an instrument calibrated in pounds of BMEP, or brake mean effective pressure. By having an indications of the

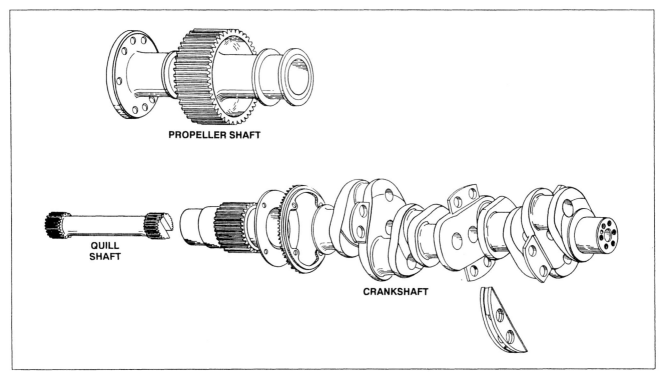

Figure 2-40. A hardened steel quill shaft may be used to drive the propeller shaft from the crankshaft and to absorb some of the torsional vibrations.

BMEP and the engine RPM, the flight engineer could know the amount of power the engine was producing.

6. Valve Operating Mechanism

a. Radial Engines

The valves on a radial engine are opened by a cam plate or cam ring, which is geared to the crank shaft. The relationship between the number of lobes on the plate and the speed of the plate relative to the crankshaft is such that all of the valves open in their proper sequence in 720° of crankshaft rotation.

When the cam plate rotates, the tappet rollers ride up onto the lobes and force the valves open by the push rods and the rocker arm.

A specific clearance must be maintained between the rocker arm and the stem tip, a clearance large enough that the valve will never be held off its seat, and yet small enough to minimize pounding which

causes excess wear. There is so much more mass in the cylinder and cylinder head than there is in the push rod that, as the cylinder heats up, it expands and the valve clearance increases. When the cam plate is timed to the engine, the valves of one cylinder, usually number one, are adjusted to the hot, or running, clearance and the cam timed to the crankshaft. Then all of the valve clearances are adjusted to the proper cold clearance, which is much smaller.

Many of the larger radial engines are designed with so much clearance between the cam plate and its bearing that the cams are called "floating cams". When the valve clearance is adjusted, the spring tension on two valves across the engine must be relieved so the cam plate can move down against the cam bearing and the clearance adjustment can be kept consistent all around the engine.

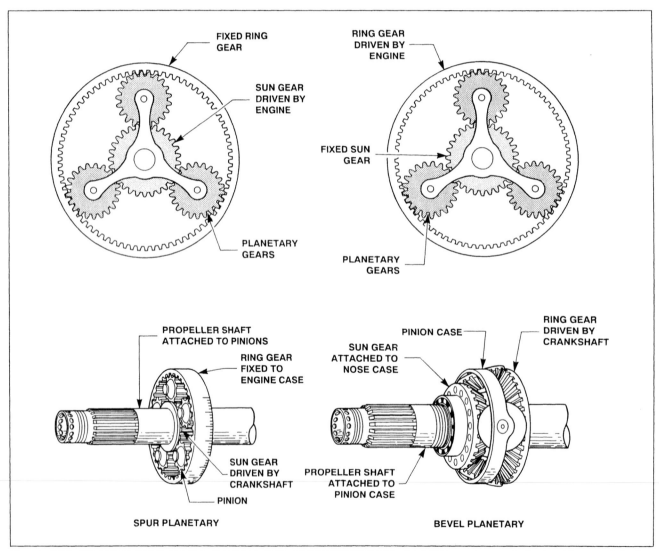

Figure 2-41. Planetary gears are the most generally used arrangement when it is necessary to transmit a great deal of power with the smallest practical frontal area.

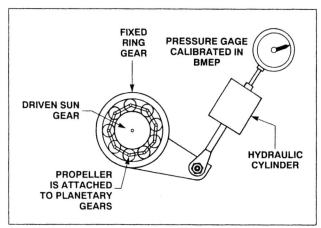

Figure 2-42. Some of the large radial engines have used a torque nose to convert the torque opposing the fixed ring gear into an indication of the BMEP.

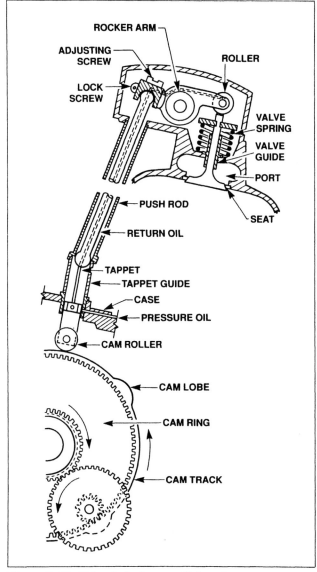

Figure 2-43. Valve train for a radial engine.

b. Horizontally Opposed Engines

The valves in horizontally opposed engines are opened in much the same manner as those in radial engines except a straight camshaft, driven at one-half engine speed, is used to time the valve opening to the crankshaft position. The cam rotates against the hardened steel tappet, or cam follower, face, and in almost all modern opposed engines, hydraulic

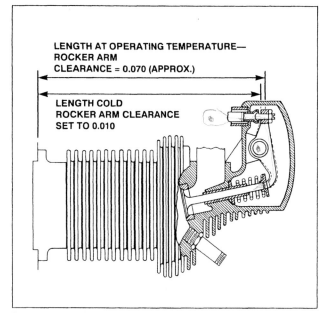

Figure 2-44. The cylinder head grows so much more than the push rob as the engine reaches operating temperature, that the rocker arm-to-valve stem clearance is much greater when the engine is hot than when the engine is cold.

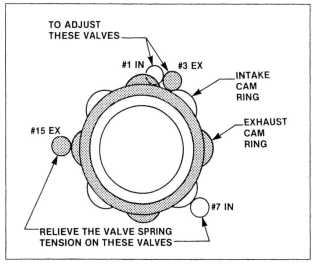

Figure 2-45. When adjusting the valves of a radial engine using a floating cam, the tension on the valve springs across the engine must be relieved so the cam rings will be pushed down against the bearing.

31

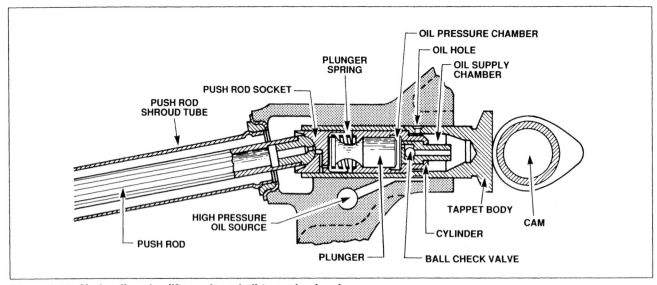

Figure 2-46. Hydraulic valve lifter using a ball-type check valve.

valve lifters are used inside the tappets to take up all of the valve clearances.

The hydraulic valve lifter consists of a plunger fitted with an extremely close tolerance into the lifter body. A spring holds the plunger away from the bottom of the body, and a check valve allows oil to enter, but will not allow it to flow out of the body. When the engine is started, oil flows through the check valve, and when the cam rotates and pushes on the cam follower, the oil in the body is trapped, and the lifter acts as a solid portion of the valve mechanism, pushing the valve open. While the valve is held open against the force of the valve spring, an extremely small and accurately calibrated amount of oil leaks out between the body and the plunger. This is so that when the valve closes, and there is no more force on the valve train, the spring in the lifter moves the plunger out, so engine oil will flow through the check valve and again fill the body. This action maintains zero clearance between the valve stem and the rocker arm, so there is no hammering, and only a minimum of wear in the valve mechanism.

7. Crankcase

The crankcase of an aircraft engine is the body that holds the engine together, and it is by the crankcase that the engine is held in the engine mounts. Horizontally opposed engine crankcases are made of two halves of cast aluminum alloy, either manufactured by sand-casting or by a method known as permamolding which produces a smoother and more dense material and allows the construction of thinner cases. Webs to support the bearings are cast into the case, and bosses are provided for mounting the cylinders.

The crankcase halves are held together by studs and through bolts which attach the cylinders, and smaller nuts and bolts around the outside of the case.

After the case halves are bolted together, the bearing cavities of the crankshaft and the camshaft are line-bored to provide almost perfect alignment.

a. Bearings

The crankshaft is supported in the crankcase by steel-backed, lead-alloy bearing inserts held in place by a tang on the bearing which fits into a groove in the bearing cavity; or, in some cases, by a dowel in the bearing cavity which fits through a hole in the insert. The camshaft rides in bearing surfaces machined in the case, but without using any inserts.

Thrust loads are carried by either a separate thrust washer or by a thrust surface on one of the

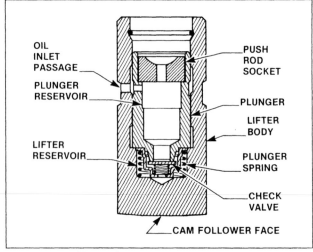

Figure 2-47. Hydraulic valve lifter using a disc-type check valve.

32

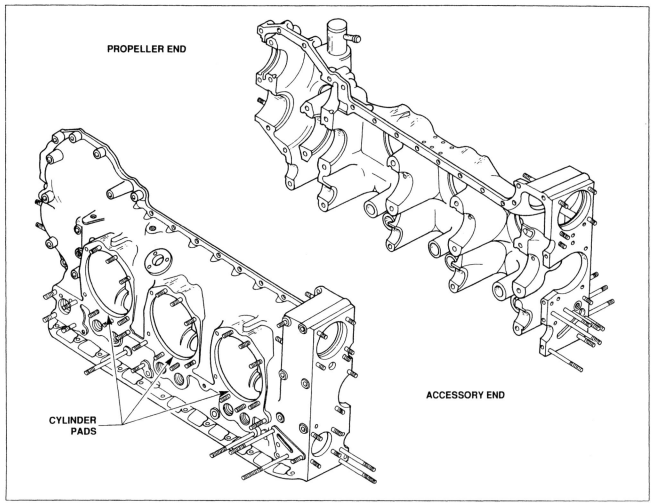

Figure 2-48. Cast aluminum aircraft engine crankcase.

PROPELLER END

ACCESSORY END

CYLINDER
PADS

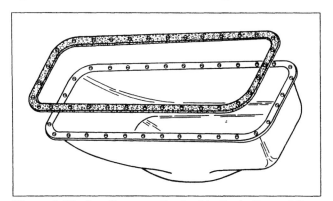

Figure 2-49. An oil sump, attached to the bottom of the engine, carries the entire engine oil supply.

main crankshaft bearings. Large radial engines use deep-groove ball bearings for carrying their thrust loads. Accessory shafts are supported in ball or roller bearings, or in some cases by needle bearings.

b. Oil Sealing Provisions

The supply of oil for the engine is carried in the crankcase, and provision must be made to prevent its leaking. The two halves of the crankcase must not only be permanently sealed against oil leakage, but the seal must not interfere with the tight fit of the bearings. Most crankcase halves are sealed with a very thin coating of non-hardening gasket compound applied to the parting surfaces with a fine silk thread embedded in the compound, extending the full length of the case. When the assembled crankshaft and camshaft are installed in the crankcase and the halves bolted together and properly torqued, the gasket material and silk thread will form an effective oil seal without interfering with the bearing fit.

The oil sump and rear case of the engine are sealed with gaskets made of either cork or an asbestos-impregnated sheeting, coated with a very light film of non-hardening gasket material. The cylinders are sealed in the crankcase by a cylinder-base packing ring. This is a form of O-ring that fits between the cylinder base skirt and the chamfered edge of the cylinder mounting pad in the crankcase.

When a rotating shaft comes through the crankcase, an oil seal must be installed that will allow the shaft

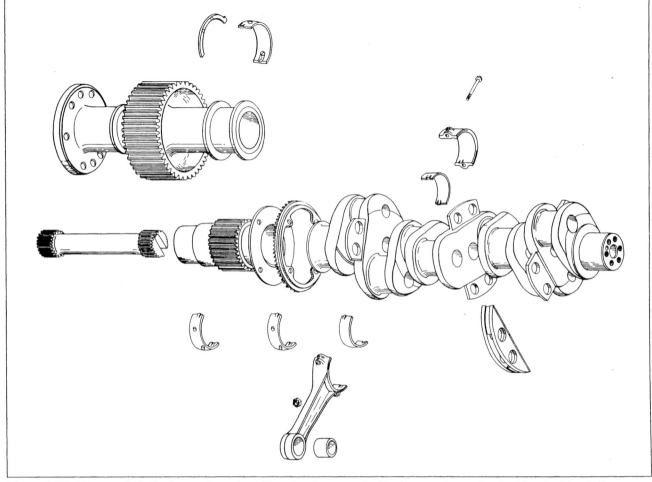

Figure 2-50. The crankshaft of a modern aircraft engine is supported in the crankcase in steel-backed lead-alloy, replaceable bearing inserts.

to rotate, yet will not allow any oil to leak out. These seals are usually made of neoprene, leather, or some form of synthetic resin such as Teflon®. The sealing surface is held against the rotating shaft by either a coil spring or a flat, leaf spring.

F. Lubrication System

1. Functions of the Oil

a. Reduces Friction

If we were to microscopically examine the surface of the metal parts of an aircraft engine, we would find that they are not perfectly smooth, but are rather made up of peaks and valleys, and when two such parts rub together, there is friction as metal is worn away from the surfaces.

To reduce this friction, a film of oil is placed between the moving parts. This oil wets the surfaces, fills in the valleys and holds the metal surfaces apart. The movement is now between layers of the oil which slide over each other with very little friction. The viscosity of an oil is its fluid friction, and the clearance between the

moving parts determines the viscosity of the oil required to prevent this film breaking away and allowing the metal to metal contact that causes wear.

b. Provides Cooling

The oil is in intimate contact with the moving parts of an aircraft engine, and it absorbs some of the heat from the combustion process. This heated oil then flows through the system into the oil cooler where the heat is given up to the outside air passing through the core of the cooler.

c. Seals and Cushions

The viscous nature of oil, that is, its ability to wet the surface it contacts, makes oil a good sealing agent between the moving parts. The oil film on the cylinder walls and around the piston increases its ability to form a tight seal in the cylinder, and the thin film of oil between the rocker arm and its bushing takes up much of the hammering shock from the valve action.

d. Protects Against Corrosion

When metal is allowed to remain uncovered in the presence of moisture or some of the chemicals that contaminate the air, rust or other surface corrosion will form. This is especially true of metal surfaces such as cylinder walls or crankshafts which have been hardened by the process of nitriding. A film of oil covering these surfaces will prevent the oxygen reacting with the metal and pitting it.

e. Cleanses

Dirt, dust, carbon, and water all enter the oil, and the ability of an oil to hold these contaminants until they can be picked up and trapped in the filter helps keep the inside of the engine clean.

2. Types of Aircraft Engine Oil

a. Straight Mineral Oil

MIL-L-6028B is a straight mineral oil that has been used for many years as the chief lubricant for aircraft engines. It has one main limitation that being its tendency to oxidize when it is exposed to elevated temperatures or when aerated.

Large amounts of carbonaceous deposits form in the turbocharger of aircraft engines because of the heat from the exhaust gases used to spin the turbine. When the engine is shut down, the turbocharger acts as a heat sink and cooks the oil in the bearings, forming carbon. Sludge also forms in this oil at the relatively low temperature of 150°F, or lower. It forms from such combustion products as partly burned fuel, water vapor, and lead compounds, and the particles unite to form a loosely linked mass which clogs oil screens and scores engine bearings.

b. Metallic-ash Detergent Oil

Certain metallic-ash-forming additives have been added to mineral oil to increase its oxidation stability. These additives were chosen to have the minimum effect in the combustion chamber and minimize

Figure 2-51. Flexible packings prevent oil leaking between a rotating shaft and the stationary crankcase.

spark plug fouling and pre-ignition, and their cleaning action in the engine loosened any carbon deposits or sludge that had formed; once these were loosened they would pass through the engine where they were prone to clog oil passages and filters.

Detergent oils were given only a limited approval by some of the engine manufacturers and, because of their limitations, have passed off the scene as an oil suitable for aircraft use.

c. Ashless Dispersant Oil

By far the most important oil in use today is ashless dispersant (AD) oil. It does not have the carbon-forming restrictions of straight mineral oil, nor does it form ash deposits as the detergent oil did.

Meeting the specification MIL-L-22851, AD oil is approved by Lycoming, Continental, Pratt and Whitney, and Franklin for use in their engines, and is the only oil used by the military services for their piston engine aircraft.

Ashless dispersant oil does not have any of the ash-forming additives, but uses additives of the dispersant type, which, instead of allowing the sludge-forming materials to join together, causes them to repel each other and stay in suspension until they can be picked up by the filters. It has been argued that these contaminants, held in suspension, will act as liquid hones and accelerate the wear of the engine parts, but this has proven untrue. It is interesting to note that most engine manufacturers recommend that new engines be operated on straight mineral oil for the first fifty hours, or at least until oil consumption stabilizes, and then switch to AD oil. The reason for this is that AD oil has so much better lubricating characteristics that it will not allow enough wear to properly seat the rings.

d. Synthetic Oil

The higher operating temperatures of modern reciprocating engines, and the lower temperature environment in which turbine aircraft operate, have in the past few years caused synthetic oil to be produced which is proving superior to mineral oils for lubrication. At the present time, this oil enjoys only limited approval for piston engine aircraft use, not because of any inherent problems, but because of lack of service experience.

The low-temperature operating characteristics of synthetic oil makes it admirable for cold weather and high altitude operation. Its viscosity at –20°F is about the same as ashless dispersant mineral oil at 0°F, and engines operated on synthetic oil have been successfully started and run at ambient temperatures as low as –40°F. This characteristic has the potential advantage in cold weather operation of eliminating the extensive engine pre-heat and the practice of draining the oil when arriving in a warm climate, making it truly an all-weather oil.

Additive synthetic oils appear to have advantages over mineral oils with respect to engine cleanliness. Oil oxidation at high temperatures, with its resultant deposits, has shown in laboratory tests to be less than that produced in either straight mineral oil or ashless dispersant oil. Synthetic oils replaced mineral oils for turbine operation when their high power and high-temperature operation became a limiting factor. These oils inhibit oxidation and thermal decomposition at temperatures well above the range of mineral oils. In the 150-hour certification tests, they have demonstrated their ability to operate satisfactorily with bulk oil temperatures as high as 245°F.

The wear characteristics of synthetic oil appear to be about the same as those of ashless dispersant oil and superior to straight mineral oil.

One of the problems with synthetic oil is its more pronounced softening effects on rubber products and resins. One manufacturer requires a more frequent replacement of inter-cylinder drain lines when synthetic oil is used, and pleated paper oil filters must be examined more closely to be sure the oil does not dissolve the resins and allow the filter to collapse. Synthetic oil is considerably more expensive than the more commonly used mineral oils, but the extended drain period for this oil appears to compensate for its higher cost. Any decision to use synthetic oil, taking into consideration cost factors, must be an individual matter based on specific operating conditions to determine whether or not the advantage of the oil justifies its increased cost.

The concept of synthetic oil for reciprocating aircraft engines is relatively new, and engine manufacturers have not as yet resolved all of the problems, nor have they issued a blanket approval for them. This does not infer any condemnation; it merely indicates that sufficient data has not been obtained, nor has field evaluation been completed.

3. Compatibility of Oils

Contrary to popular opinion, oils within their basic categories are compatible. All straight mineral oils are physically compatible with each other, and it is extremely doubtful if engine performance or cleanliness would be compromised by mixing several brands of oil.

All ashless dispersant mineral oils meeting MIL-L-22851 specifications are physically compatible, and one brand may be added to an engine operating on another brand without destroying the overall cleanliness of the engine. This was proven by the military when no performance or cleanliness deterioration resulted from operating between suppliers having different brands.

Ashless dispersant oils are also compatible with straight mineral oil. The airlines proved this while converting to the ashless dispersant type. In the transition, various amounts of straight mineral oil and ashless dispersant oil were mixed, and it was found that there were no adverse side effects other than a reduction in the degree of the advantages from operating with all AD oil.

	AN 1065 SAE 30 Aviation 65	AN 1080 SAE 40 Aviation 80	AN 1100 SAE 50 Aviation 100	AN 1120 SAE 60 Aviation 120
Viscosity				
SSU @ 100°F	443.0	676.0	1,124.0	1,530.0
SSU @ 130°F	215.0	310.0	480.0	630.0
SSU @ 210°F	65.4	79.2	103.0	123.2
Viscosity Index	116	112	108	107
Gravity, °API	29.0	27.5	27.4	27.1
Color, ASTM	1.5	4.5	4.5	5.5
Pour Point, °F	–20	–15	–10	–10
Pour Point, Diluted, °F	–70	–70	–70	–50
Flash Point, °F	450	465	515	520
Carbon Residue %W(R)	0.11	0.23	0.23	0.40

Figure 2-52. Characteristics of aircraft engine lubricating oil.

At the present time, because of lack of experience, it is not advisable to add synthetic oil to engines operating on straight mineral oil, or on ashless dispersant oil. Before changing from either of the mineral oils to synthetic, be sure to follow in detail the procedures for flushing and draining established by the synthetic oil manufacturer.

4. Engine Oil Ratings

a. Viscosity

The viscosity, or fluid friction, of an engine oil is one of its more important ratings, and is measured by a rather elaborate laboratory instrument known as a Saybolt Universal Viscosimeter. The number of seconds required for 60 cu. cm. of the oil to flow through an extremely accurately calibrated orifice, at a specified temperature, is known as the SSU or Saybolt Second Universal, viscosity.

A convenient grading of engine oil for aviation use is done by rounding off the SSU numbers for 210°F, as we see in figure 2-52. The Society of Automotive Engineers (SAE) and the military have different numbering systems for the same grade of oil, but the numbers all relate to the SSU viscosity.

b. Viscosity Index

Viscosity index is a measure of the change in viscosity of an oil with a given change in temperature. The index itself is based on the viscosity changes with temperature of two reference oils, one rated 100, and the other zero. The less change there is in viscosity for a given temperature change, the higher the viscosity index.

c. Gravity, API

The American Petroleum Institute (API) has formulated a measurement of the specific gravity of petroleum products which is an expansion of the regular specific gravity scale. An API gravity of 10.0 is the same as a specific gravity of 1.00, and API 63.8 corresponds to a specific gravity of 0.7245; a conversion chart can be used to relate any API number between these values to the corresponding specific gravity.

d. Color

The color of an oil is rated by comparing its color with an ASTM (American Society for Testing and Materials) color chart. One, on the chart, is pure white, and eight is a darker red than claret wine.

e. Pour point

The pour point of an oil is the lowest temperature at which the oil will pour without disturbance.

f. Flash point

The flash point is that temperature to which oil must be raised before it will momentarily flash, but not sustain combustion when a small flame is passed above its surface.

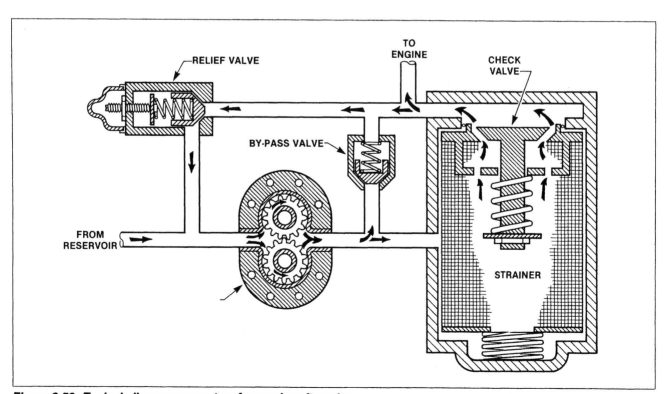

Figure 2-53. Typical oil pressure system for an aircraft engine.

g. Carbon Residue

A given amount of oil is placed in a stainless steel receptacle and heated to a specifically high temperature until it is evaporated. The container is weighed before and after the test, and the amount of carbon residue left in the container is expressed in this rating as a percent of the weight of the sampled material.

5. Lubrication Systems

a. Dry Sump System

Radial engines are not built in such a way that the lubricating oil can be carried in the engine crankcase, and it was this engine design that brought about what we know as a dry sump system. The oil supply is carried in an external tank attached to the engine mount with clamps, or built into the structure behind the firewall.

Oil is fed from the tank into the engine by gravity, and in this supply line at its lowest point is an oil drain Y-valve. This valve may be opened to drain the entire system, and is safetied in the shut position for normal operation. An oil temperature probe is inserted in this line near the point it enters the engine to measure the temperature of the inlet oil. Inlet temperature measures the effectiveness of the cooling system, as well as giving an indication of abnormal temperatures that may exist in the engine.

From the inlet line, the oil is picked up by a gear-type pressure pump and passed to the strainer. Most of these strainers use screens with two sizes of mesh; the larger mesh provides the mechanical shape and strength, while the smaller screen does the filtering. A relief valve is built into the strainer that will open if the screen should become so clogged that no oil can pass through it, and a built-in check valve prevents oil draining into the engine when the pump is not operating. The filtered oil flows through drilled passages to lubricate the crankshaft bearings and the cam bearings and through the hollow propeller shaft to provide oil for hydraulic propeller control. A spring-loaded oil pressure relief valve dumps the oil back into the inlet of the pump when the pressure exceeds that for which it is set.

Oil to lubricate the valve mechanism flows into the rocker boxes through hollow push rods and drains back either through inter-cylinder drain lines, or the push rod housings.

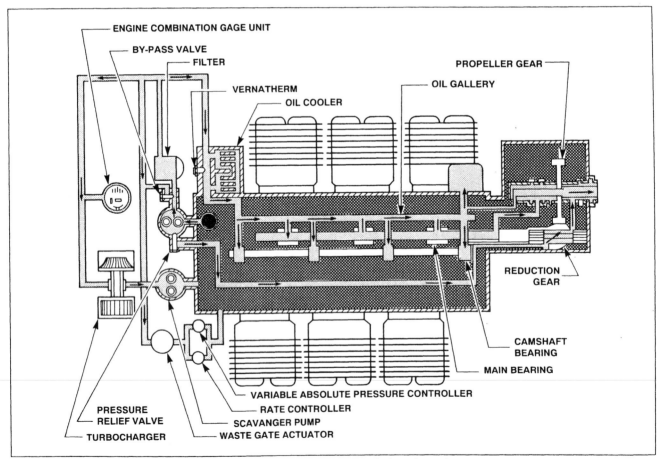

Figure 2-54. Lubrication system for a modern horizontally opposed aircraft engine.

38

After the oil has circulated through all of the engine, it drains by gravity and collects in one or more sumps, which are small compartments connected by oil pickup tubes to the scavenger pump. A scavenger pump is similar to a pressure pump, but is considerably larger because the oil being scavenged is hot and has air entrapped, causing its volume to be greater than that delivered by the pressure pump.

From the scavenger pump, the oil leaves the engine and is directed through the oil cooler and then back into the tank.

b. Wet Sump System

Most modern aircraft engines use a wet sump system in which all of the oil is carried in a sump which is part of the engine itself. The oil is picked up through an oil suction tube, by the gear-type pump. From the pump exit, the oil is directed to the oil filter chamber, and if the filter should become clogged, a spring-loaded bypass valve will open, permitting oil to flow directly from the pump to the oil filter outlet. From here, a passage leads to the adjustable oil pressure relief valve, and any time the pressure exceeds that for which the valve is set, the excess bypasses back to the inlet side of the pump.

Oil at the regulated pressure flows into the oil cooler, entering through the temperature control valve. If the temperature is below that for which the valve is set, the oil goes directly into the oil gallery in the engine crankcase; but if it is hotter, it is directed through the core of the cooler. This valve keeps the oil temperature within the range specified for the engine operation.

The oil flows through drilled passages to the front of the engine, providing lubrication for the crankshaft, camshaft, and propeller shaft bearings, the valve mechanism, and the propeller governor.

The gear-type pump in the propeller governor boosts the engine oil pressure and directs it back into the crankcase through drilled passages, through the hollow propeller shaft, and into the propeller to control the pitch.

The accessory drives are lubricated by oil fed into their bearings through drilled passages, and the gears are cooled and lubricated by a spray of oil in the accessory case at the rear of the engine.

The cylinder walls are lubricated, and the pistons are cooled by oil from around the crankshaft main bearings which is squirted in a continuous stream against the inner dome of the piston. This oil, after it has lubricated the pistons and cylinders, drains down into the sump.

The turbocharger is lubricated by oil from the crankcase gallery, which, after it has lubricated these critical bearings, drains into an oil separator, which removes the air, and the oil is picked up and returned into the engine sump by a gear-type scavenger pump.

6. Lubrication System Components

a. Pumps

(1) Gear-type

Two spur gears, one driven by the engine and the other by the drive gear, rotate in a close-fitting housing. As the teeth unmesh on the inlet side of the pump, the volume of the cavity increases, lowering the pressure and drawing oil into the pump. This oil is trapped between the teeth and the housing and carried around the outside of the gears to the outlet side. As the teeth of the gears mesh, the volume of the cavity decreases, and the oil is forced out of the pump into the drilled passages in the engine crankcase.

A gear pump is a constant displacement pump, meaning that a specific amount of fluid is moved each time the pump rotates, and provision must be made to relieve the excess pressure so the system pressure can remain constant as the pump speed varies.

(2) Gerotor-type

Another form of constant displacement pump used for moving lubricating oil through a reciprocating engine is the gerotor pump, a special form of gear pump. In the illustration of figure 2-56, we see a 6-tooth spur gear driven from an engine accessory drive. This gear rides inside a rotor which rotates freely in the housing. A seven-tooth internal gear is cut inside the rotor, and

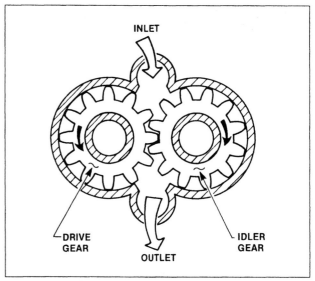

Figure 2-55. Gear-type oil pump.

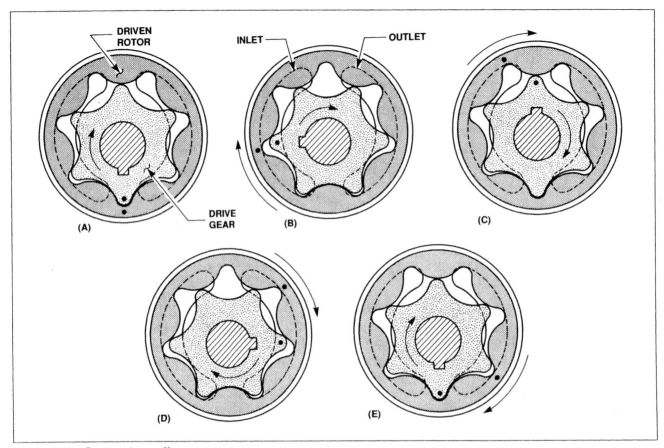

Figure 2-56. Gerotor-type oil pump.

as the drive gear is turned, it rotates the rotor. In A, the two marked teeth are meshed and there is a minimum of space between them. As the two gears rotate, the volume between the teeth increases as you notice the marked teeth in A, B, and C. A plate with two kidney-shaped openings covers the gears, forming a seal for their ends. As the gears rotate beneath the inlet port, the volume between the teeth continually increases, and as they rotate beneath the outlet port, as seen in D and E, the volume decreases, moving the fluid out into the system. Gerotor pumps may be designed to pump relatively large volumes of fluid without their having to be excessively thick.

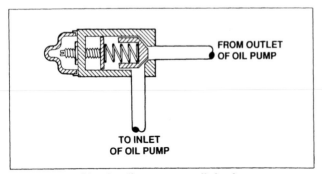

Figure 2-57. Simple oil pressure relief valve.

b. Relief Valves

(1) Simple Relief Valve

Almost all pumps used in aircraft engine lubrication systems are of the constant-displacement type, and provision must be made to relieve some of the oil back to the inlet of the pump to maintain the pressure constant as the engine RPM changes. Figure 2-58 shows the operating principle of a simple pressure relief valve. Oil from the discharge side of the pump flows into the lubricating passage, which is connected by a spring-loaded valve back into the inlet side of the pump. As long as the pressure is below that for which the relief valve is set, the valve remains on its seat; but when the pressure rises, the valve will move off its seat, and oil will return to the inlet side of the pump, holding the system pressure constant. An adjustable screw, locked with a jamnut and covered with a protective cap, is used to change the oil pressure. Loosening the jamnut and turning the adjustment screw clockwise raises the oil pressure.

(2) Compensated Relief Valve

Some of the larger engines require a very high oil pressure to force the cold oil through all of the bearings for starting. But when the oil warms up, and becomes thinner, this high pressure would

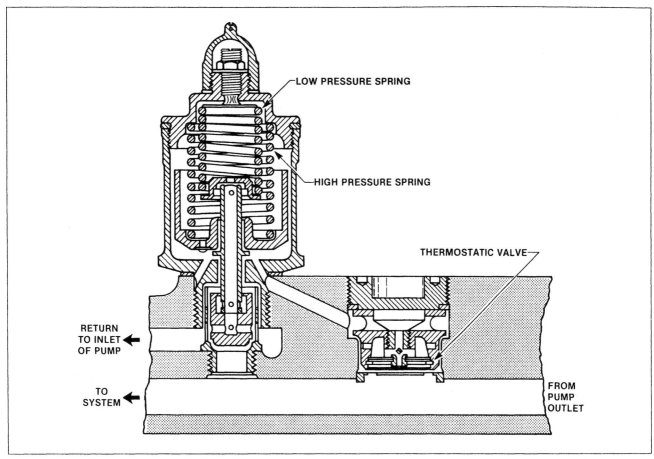

Figure 2-58. Compensated oil pressure relief valve allows the pressure to be high enough to force the cold oil through all of the passages, but to decrease it into the proper operational range when the oil warms up.

cause an excessive oil consumption, so a compensated pressure relief valve is used to hold the pressure high until the oil warms up, then automatically lowers it to the normal operating range. This is done by using two springs to hold the valve on its seat when the oil is cold, and when the oil warms up, a thermostatically operated valve opens a passage and allows oil to flow beneath a piston, relieving the force on the high-pressure spring. Normal operating pressure is maintained by the force of the low-pressure spring along.

c. Oil Coolers

Much of the heat released from the fuel is picked up by the oil, so some provision must be made to transfer part of it into the air. This transfer is done in a radiator, or an oil cooler.

Depending on the design of the lubricating system, the cooler may be either between the oil filter and the bearing, or it may be in the return line between the scavenger pump and the oil tank. In either case, a thermostatic control valve allows the oil to bypass the core of the cooler when it is cold, and as it warms up, the valve forces the oil through the core so the excess heat may be picked up by the air.

d. Oil Reservoir

Dry sump engines carry their oil in an external reservoir which must have a capacity compatible with the fuel carried, plus a margin that will provide for adequate circulation and cooling.

Engines operated in cold weather must have some provision for thinning the oil for starting. In extreme cases, the oil may be drained while it is still hot from the last flight of the day, and warmed the next morning before it is put back into the oil tank; but a much simpler method for thinning the oil is oil dilution.

Before the engine is shut down, and while the oil is still hot, fuel from the carburetor is directed into the oil at the inlet Y-valve where it mixes with the oil, diluting it. Rather than having to dilute the entire oil supply, engines equipped for dilution have hoppers in the oil tank. A hopper is a loose-fitting cylinder, often with holes in its side, between the oil return to the tank and the line to the pump. The oil is diluted for the proper length of time for the temperature anticipated for starting, and this diluted oil fills the hopper. When the engine is started, the diluted oil in the hopper is used first and circulated through the engine until it is warm, and then the oil surrounding the hopper flows in through the holes and mixes with the oil circulating through the engine. When the oil warms up, the gasoline evaporates from it and leaves

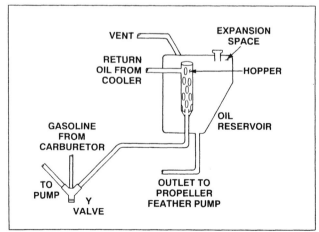

OIL
OUT

TEMPERATURE
CONTROL VALVE

OIL
IN

**WHEN THE OIL IS COLD, IT FLOWS AROUND
THE CORE OF THE COOLER.
(A)**

**WHEN THE OIL IS HOT, THE TEMPERATURE CONTROL
VALVE SHUTS OFF THE PASSAGE IN THE SHELL OF
THE COOLER, AND THE OIL MUST PASS THROUGH
THE CORE, WHERE IT TRANSFERS ITS HEAT TO
THE AIR PASSING THROUGH THE COOLER.
(B)**

THERMOSTATIC
OIL
TEMPERATURE
CONTROL
VALVE

OIL
INLET
LINE

COOLER

OIL OUTLET LINE

**TYPICAL OIL COOLER INSTALLATION
(C)**

Figure 2-59. Engine oil cooler.

the engine through the crankcase breather, and the oil returns to its original condition.

Oil tanks used with engines equipped with hydromatic propellers feed the pressure pump from a standpipe which sticks into the tank for a few inches. The tank outlet to the propeller feathering pump is taken from the bottom of the tank, and with this arrangement, if an oil line should break and all of the oil be pumped overboard by the engine pump, there will still be enough oil in the tank to feather the propeller.

e. Filters

Solid contaminants and sludge pumped through an aircraft engine lubricating system can clog the oil passages and damage the bearing surfaces, so provision must be made to remove as much of it as possible. There are two ways this may be done: by a full-flow filter, through which all of the engine oil must flow each time it circulates through the engine, or with a bypass filter that filters only a portion of the oil each time it circulates, but which eventually gets all of the oil. Bypass filters can be made finer because, if they should plug and prevent oil flow, they will not deprive the engine of its oil.

The majority of filters used in modern aircraft engines are of the full-flow type, and there are four methods of filtration used:

(1) Edge Filtration

A long, flat spiral with a wedge-shaped cross section forms the filtering element for an edge-type filter. The oil flows from the outside of the element into its center, and any contaminants collect on the outside where the slots are the smallest. Since the cross section of the spiral is wedge-shaped, if a contaminant should be small enough to pass through, it will not clog the filter. Some edge-type filters are cleaned with a built-in blade that is

VENT

EXPANSION
SPACE

RETURN
OIL FROM
COOLER

HOPPER

GASOLINE
FROM
CARBURETOR

OIL
RESERVOIR

TO
PUMP

Y
VALVE

OUTLET TO
PROPELLER
FEATHER PUMP

Figure 2-60. Hopper-type oil reservoir.

rotated to scrape off the contaminants that collect on the outside.

A Cuno-type filter is another form of edge filter, whose filtering element consists of a stack of very thin metal rings with fixed radial spacers between them. The rings are attached to a handle which sticks through the case, and may be turned during an inspection. Contaminants collect on the outside of the stack of rings, and when the stack is rotated, the spacers scrape all of the residue out and it falls into a compartment in the bottom of the filter housing where it can be removed on a regular maintenance inspection.

(2) Depth Filtration

Depth filters consist of a matrix of fibers that are closely packed to a depth of about one inch. The oil flows through this mat and the contaminants are trapped in the fibers. Depth-type filters are more efficient than the edge-type because they can hold so much more contamination, but because of the non-uniformity of the matted material it is possible for the high-pressure oil flowing through it to find a weak spot and channel through, causing the filter to lose a great deal of its effectiveness.

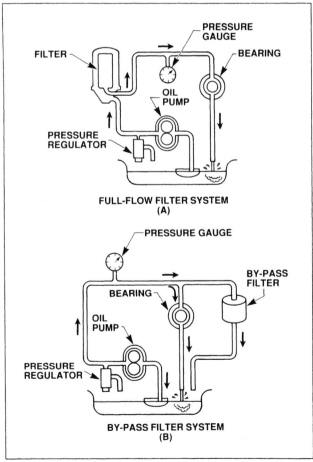

Figure 2-61   Oil filter systems.

(3) Surface Filtration

Standard equipment in almost all aircraft engines is a woven wire-mesh oil filter that is useful for trapping some of the larger contaminants that flow through the engine. Particles of contamination

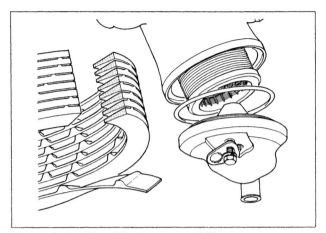

Figure 2-62. Edge filtration-type oil filter.

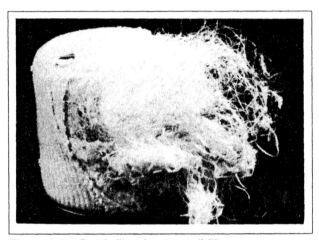

Figure 2-63. Depth filtration-type oil filter.

Figure 2-64. Screen-type surface filtration oil filter.

larger than the size of the screen openings are trapped on the surface, and the smaller particles pass through. To get better filtering, the size of the wire mesh must be decreased, and when this is done, the cleaning process becomes quite complex and the advantage of simplicity is lost. As a result, this type of filter is often supplemented by a more efficient type.

(4) Semi-depth Filtration

The most popular filter in common use for general aviation aircraft is a disposable, semi-depth filter made up of resin-impregnated fibers, figure 2-65(A), formed into a long sheet, and folded in a pleated fashion, figure 2-65(B). All of the oil that flows in the system must pass through this element, and in doing so, the contaminants are trapped within its fibers. There is so much uniform surface area that there is little tendency for the oil to channel through. The pleated material is assembled around a sheet steel core, figure 2-65(D), and fitted either in a heavy sheet steel case which is an integral part of the filter, or made so it may be installed in a housing which is part of the engine.

When this type of filter is removed from the engine at inspection time, it is cut open and inspected for the presence of any metal particles that might indicate an impending engine failure. Sealed, spin-on-type filters are opened with a special roller-type can cutter that removes the top of the container without introducing any metal particles that might make the examination more difficult.

f. Instrumentation

(1) Pressure Measurement

Oil pressure is measured at the outlet of the engine driven pump and is indicated on a gage on the pilot's instrument panel. To prevent gage fluctuations and keep the loss of oil to a minimum if the line to the instrument should break, a small orifice, about a number 60 drill size, is installed in the fitting where the gage line attaches to the oil pressure gallery in the engine.

(2) Temperature Measurement

The oil temperature measurement read on the instrument panel is the temperature of the oil entering the engine. On dry sump engines, the temperature pick-up bulb is located in a special fitting in the line between the oil drain Y-valve and the pump, and on dry sump engines, it is installed inside the oil screen immediately after the pump. Temperature may be measured electrically, by measuring the change in resistance of a special temperature probe, or mechanically by measuring the pressure of a gas

sealed in a bulb, held in the oil. The pressure of the gas varies proportionally to its temperature.

G. Cooling Systems

1. Liquid Cooling System

As late as World War II, some aircraft engine were cooled by transferring the excess heat into liquid, and then piping it into a radiator where the airflow could absorb the heat. Water was first used as a coolant, but since it boils away at altitude, without absorbing very much heat, another medium for absorbing the heat had to be found. Ethylene glycol, commonly called Prestone®, was used in a sealed cooling system which allowed the engine operating temperatures to be high enough to satisfy the engine's requirements; and at the same time, the sealed system prevented the coolant boiling away. The weight of the cooling system and the coolant, along with the complexity of all of the pumps and plumbing has made liquid cooling unpopular, and for practical purposes today, all currently produced aircraft engines are air-cooled.

2. Air-cooling System

The cylinders of an air-cooled engine are finned to increase the area exposed to the air. Thin fins are machined onto the alloy steel barrel, and deep fins are cast into the aluminum alloy head. In figure 2-66, we see that the area around the exhaust valve has deeper fins than that around the intake valve because of the greater amount of heat in the head on that side.

The cooling air enters the cowling through the air inlets at the front of the engine and is directed by baffles through the cylinder fins. Thin sheet metal inter-cylinder baffles aid in forcing the air through the fins so the maximum amount of heat can be extracted from the cylinders.

The amount of air that can flow through the cylinders is determined by the air pressure differential between the top of the engine and the space below the engine. On most high-output engines, this pressure differential may be varied with cowl flaps, figure 2-68. When the engine is on the ground and the airflow is restricted, the cowl flaps are opened so the blast from the propeller will create a low pressure below the engine and pull more air through the fins. In flight, the cowl flaps are closed as there is enough ram air through the cowling for adequate cooling.

Augmenter tubes are used on some engines to augment, or increase, the airflow through the engine for cooling. The exhaust is directed into a stainless steel, venturi-shaped tube where its high-velocity

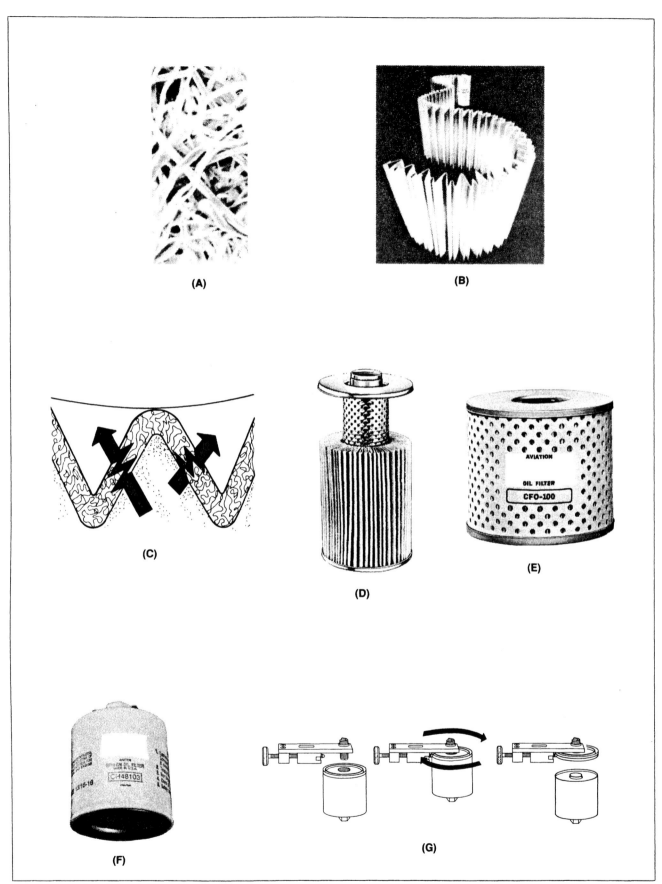

Figure 2-65. Semi-depth filtration.

flow creates a low pressure. This low pressure below the engine accelerates the flow of air through the cylinder fins, and increases the effectiveness of the cooling system.

H. Carburetion System

For an engine to develop power, there must be the proper amount of fuel supplied to the cylinder, and this fuel must be mixed with the proper amount of air. This metering may be done with a simple float-type carburetor, with a pressure carburetor, or with a fuel injection system. All of these systems do the same thing: weigh the air entering the engine, meter into this air the proper amount of fuel, mix the two,

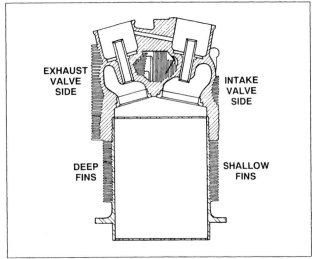

Figure 2-66. The cooling fins on an air-cooled cylinder are deeper on the side having the exhaust valve in order to carry away the excessive heat on that side.

Figure 2-67. The engine compartment is baffled to force the cooling air between the fins on the cylinders so the maximum amount of heat will be carried away.

and then distribute the fuel-air mixture evenly to all of the cylinders.

All of these functions are necessary for proper engine operation, and some types of fuel metering systems are more efficient than others. The principle of fuel metering is covered in detail in the IAP, Inc. training manual entitled *Aircraft Fuel Metering Systems*, but here we will discuss the basic functions of the system in order to better understand the way they allow an aircraft engine to develop its power.

1. Main Metering System

The basic difference between a float carburetor and a pressure carburetor is in the way the metering force is generated. In the float carburetor of figure 2-70, the air flowing into the engine passes through a venturi. In this particular carburetor, two venturis are used to increase the pressure drop. The air is speeded up as it passes through the venturis and this increase in its velocity decreases its pressure.

The fuel is held at a constant level below the lip of the discharge nozzle by a float-operated needle valve, and atmospheric air pressure acts on the fuel in the float bowl. The more air there is flowing into the engine, the lower the pressure will be at the discharge nozzle and the more fuel there will be forced into the airstream.

A butterfly-type air valve is located in the main air passage, and is controlled from the cockpit by the throttle. The more the throttle is opened, the more air can flow into the engine, and the more fuel will be mixed with the air. The main metering jet is a fixed orifice that determines the amount of fuel that is allowed to flow for any given pressure drop.

Pressure carburetors, such as the typical example of figure 2-71, measure the amount of airflow by the

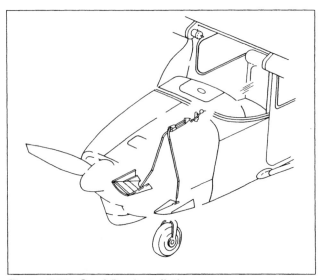

Figure 2-68. Cowl flap installation.

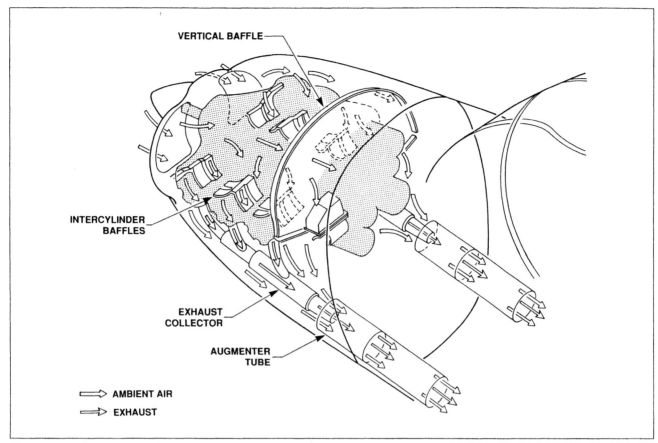

Figure 2-69. Augmenter tubes.

pressure difference between ram air pressure, picked up by the impact air passage, and the low pressure developed in the venturi. These two pressures acting on a single diaphragm move it proportionally to the amount of air flowing into the engine, and the diaphragm moves a poppet valve in series with the main metering jet so the fuel flow is proportional to the amount of air taken into the engine.

2. Mixture Control System

The amount of energy released from the fuel is proportional to the *weight* of the air mixed with the fuel. And as the airplane goes up in altitude, the air becomes less dense, meaning there are fewer pounds of air per cubic foot; so to prevent the fuel-air mixture ratio becoming rich, the amount of fuel must be decreased independent of the airflow. This is done with the mixture control, which on the simple float carburetor of figure 2-72 is simply a valve between the float bowl and the main metering jet. The valve may be partially closed to lean the mixture, or completely closed to shut off all flow of fuel to the engine. This is called the idle cut-off position.

The pressure carburetor of figure 2-71 uses a needle valve between the low- and high-pressure air chamber that is moved to vary the pressure differential between the two chambers. For a rich mixture, the needle is closed, so the pressure *difference* will be maximum; to lean the mixture, the needle is pulled out, decreasing the pressure differential so the poppet valve will allow less fuel to flow.

3. Idle System

The main metering system is effective when the engine speed is above about 1,000 RPM, and below this, there is insufficient airflow to create enough pressure drop to pull the fuel out with adequate uniformity. An idle system is used that acts as a totally independent fuel metering system. Fuel is drawn from the float bowl and metered through an idle metering jet, and air is mixed with this fuel and the resulting emulsion drawn up to the idle discharge holes at the edge of the throttle air valve. This valve is almost closed when the engine is idling, and all of the air that flows into the engine must pass around its edges. Because of this restriction, the air has a high velocity, and the pressure at the edge of the valve is low, pulling the fuel out into the air as a spray through the discharge holes.

The idle RPM of the engine is adjusted by varying the amount of air allowed to flow past the throttle valve, and the smoothness of the idling is determined

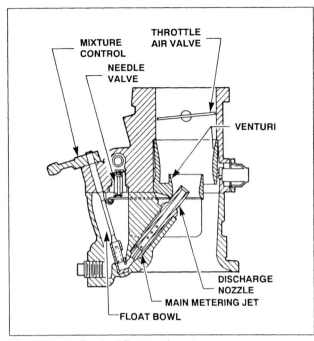

Figure 2-70. *Typical float carburetor.*

by the amount of fuel metered into this air. This is done by the idle mixture adjustment needle valve.

The idling adjustment with a pressure carburetor is made by controlling the amount of air that can flow past the throttle valve, by varying its closed position. Since the fuel is under pressure, the idle mixture is adjusted by changing the size of the orifice through which the fuel flows to the discharge nozzle.

4. Acceleration System

When the engine is idling, the throttle air valve is closed, and all the air that enters the engine must flow around its edge, and all of the fuel is discharged from the idling system. When the throttle is suddenly opened, there is a definite lag between the time the idle system ceases to function and the time there is enough airflow for the main system to begin to meter. This lag can cause such a momentarily lean mixture that the engine will hesitate, and this hesitation usually occurs at the most inopportune time. To prevent this hesitation, an acceleration system is

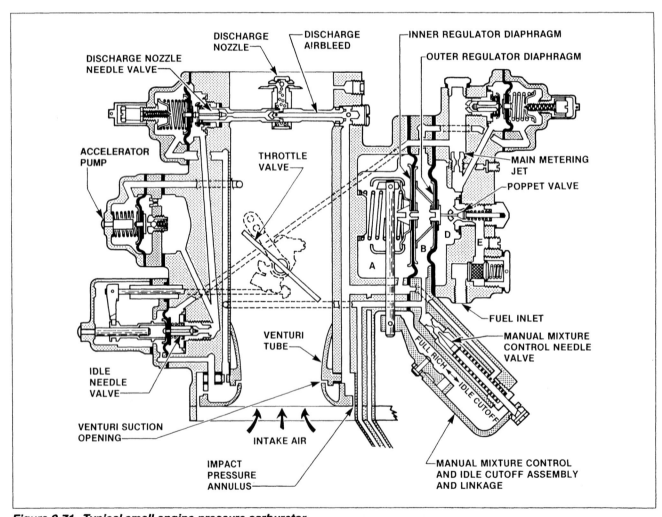

Figure 2-71. *Typical small engine pressure carburetor.*

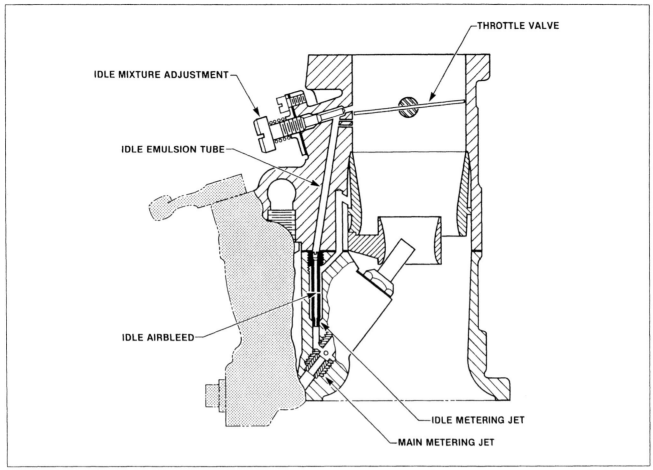

Figure 2-72. The idle system in a float carburetor is actually a separate metering system that is operational when the airflow through the engine is not sufficient to actuate the main metering system.

built into the carburetor to provide a momentarily rich mixture during this transition. In figure 2-73, we see a typical acceleration system used in a float carburetor. When the throttle is opened suddenly, the acceleration pump is pushed down, and the fuel in the pump chamber is forced out into the airstream. There is enough fuel to provide the rich mixture needed to prevent the hesitation during acceleration.

5. Power Enrichment System

Aircraft engines are built so light for the amount of power they develop, that it is easy for them to be damaged by overheating. To prevent this, many engines have a power enrichment system which provides an excessively rich mixture for full power, so the additional fuel will absorb some of the heat energy. Enriching the mixture at full throttle will decrease the power slightly, but this loss is necessary, as detonation is quite likely to occur if the excess heat is not removed.

Large, high-powered reciprocating engines during World War II used an anti-detonation injection system,

which, under full power conditions, automatically leaned the fuel-air mixture until peak power was developed, and then a water alcohol mixture was injected into the cylinders to remove the excess heat. When the water-alcohol mixture was exhausted, the mixture automatically reverted back to its enriched condition in order to prevent detonation.

6. Fuel Injection Systems

Fuel injection systems are used on many of the more powerful reciprocating engines to improve the uniformity of the fuel-air mixture distribution over that possible with a carburetor. The fuel injection system does exactly the same thing as a carburetor, and the main difference between the two is the place where the fuel is actually discharged. In a carburetor, the fuel is discharged into the air in the carburetor itself, and the fuel-air mixture travels through the induction pipes to the intake valve in the cylinder. In a fuel injection system, only the air travels through the induction system and the fuel is discharged into the air as it enters the intake valve of the cylinder. This provides a more uniform fuel-air

49

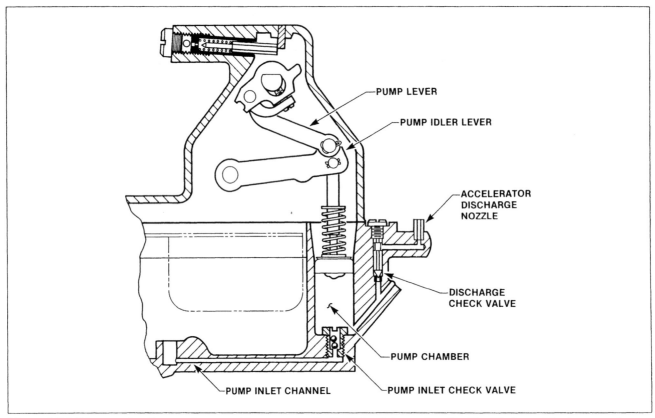

Figure 2-73. *The acceleration system provides a momentarily rich mixture when the throttle is suddenly opened.*

mixture to the cylinders and decreases the distribution difference caused by the varying lengths of the induction pipes.

I. Induction System

Aircraft engines are air breathing, and there must be sufficient airflow into the engine to provide the oxygen needed to mix with the hydrocarbon fuel so it can release its energy. The air that enters the engine must be clean, and it must be cool so its density will be high, yet not cold enough for ice to form in the induction system.

1. Air Filters

The propeller stirs up a lot of dirt on the ground, and the pumping action of the pistons causes the carburetor or fuel injection system to draw in huge volumes of this contaminated air. Sand and dust storms can fill the air with sharp-edged particles to as high as 10,000-15,000 ft. To prevent the abrasive action of the dust cutting away on the inside of the engine, effective filters must be used to trap all of it before it can enter the engine.

There are three basic types of filters used for modern aircraft engines: the flocked screen wire, the disposable paper, and the glycol-impregnated polyurethane foam.

Figure 2-74. *It is extremely important that all of the air entering the engine is filtered to remove any contaminants that could cause engine wear.*

The flocked screen wire filter is the oldest of the three types and is used on many of the smaller engines. A double layer of pleated screen wire is covered with flock, a material composed of pulverized fibers glued to the wire. These filters may be

50

cleaned by washing them in varsol, soaking them in a mixture of varsol and engine oil, and then allowing them to drip dry. Enough of the oil will remain in the flock to trap dust particles from the air and hold them until the filter can be washed and re-oiled.

Paper filters, similar to those used in automobile engines, are gaining popularity as standard equipment on many of the newer aircraft. Air passes through the porous filter element, but any dust and sand particles are trapped on its surface. Paper filters may be cleaned by blowing all of the dust out of them in the direction opposite the normal airflow and washing them in a mild soap and water solution, then allowing them to dry.

The most effective filter, and the only one used for the most severe dust conditions, is the polyurethane foam filter which is impregnated with a glycol solution. The glycol gives these filters an affinity for dust, and at the time of recommended filter change, the foam element is removed and discarded and a new one installed. It is not recommended that these filters be cleaned.

2. Carburetor Heat

A great deal of heat is required to change liquid fuel into fuel vapor in the carburetor, and this heat comes from the air and from the metal of the carburetor, causing it to act as a very effective refrigerator, lowering the temperature of the incoming air enough to condense this moisture and freeze it. All certificated aircraft must provide a means of heating the air entering the carburetor by 90°F at 75% power for sea level engines, and 120°F for altitude engines. This heating is done by taking air from the engine compartment and conducting it around the exhaust pipes or the muffler, and then directing it into the carburetor.

A carburetor heat control in the cockpit allows the pilot to select either cold, filtered air, or hot, unfiltered air for the carburetor.

The use of carburetor heat at high power settings must be avoided because of the probability of detonation. If the air is heated before it is compressed in the cylinder, it will reach its critical temperature before its energy is extracted, and it will explode rather than burn evenly.

Fuel injected engines do not have the problem with carburetor ice that is common with the float carburetor; but there is still the probability that when flying through freezing rain or through super-cooled clouds ice will form on the air filter and block the flow of air into the engine. To prevent this happening, an alternate air source is provided which allows the pilot to shift the air inlet from the nose of the engine to a point inside the cowling where there is no danger of the inlet freezing over. When either carburetor heat or alternate air is used, the air that enters the engine is not filtered, and there is danger of sand and dust entering the engine.

3. Supercharging

The power developed by an aircraft engine depends on the amount of energy in the fuel that can be converted into useful work. As the airplane goes up in altitude, the air becomes less dense and there are fewer pounds of air to combine with the fuel, and so the power decreases steadily. This may be seen from the lower curve in figure 2-77 which shows that rated power can be developed only at sea level.

It is possible for an engine manufacturer to turn out an engine capable of producing more horsepower than it will develop with normal aspiration. In this case, a geared internal supercharger may be installed to raise the pressure of the air entering the cylinders and increase the volumetric efficiency.

A typical internal supercharger system consists of a small centrifugal impeller, gear-driven from the accessory drive train somewhere between 10 and 12 times the crankshaft speed.

Air flows into the engine through the normal air filter and the carburetor, or body of the fuel injection system. It is then compressed by the centrifugal impeller and delivered to the intake valves. With a normally aspirated engine, it is not possible to have a manifold pressure higher than the existing atmospheric pressure; but with a supercharged engine, the manifold pressure may be increased to any amount the engine manufacturer will allow. Forty to forty-five inches of mercury absolute, are common values for the takeoff manifold pressure of some supercharged engines.

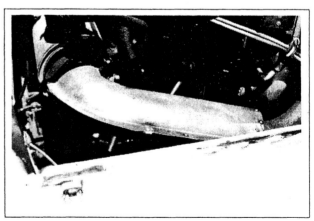

Figure 2-75. Air is passed through sheet metal shrouds around the exhaust pipes to provide heated air to prevent carburetor ice.

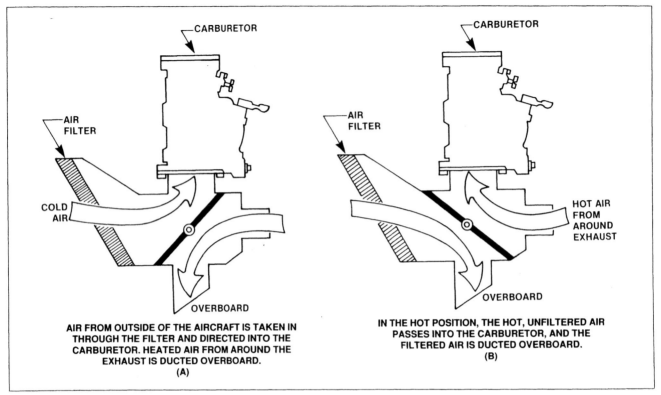

Figure 2-76. *Carburetor heat box.*

An engine with a single-speed, geared super-charger is called a ground-boosted engine. This means that the sea level rated horsepower has been increased by the use of the supercharger, but the power will decrease from this value as the airplane goes up in altitude. This is illustrated by the center curve of figure 2-77.

There have been a number of ways used to allow an engine to maintain sea level horsepower up to some specified altitude; one of these ways is the 2-speed supercharger. Takeoff and low-altitude flying is done with the supercharger operating with a low-blower ratio of about 7:1; and then, at altitude when the power has dropped by a given amount, the blower can be shifted to a higher ratio, around 9:1,

which increases the density of the air enough to provide sea level horsepower at this altitude. Another method was to use two stages of super-charging; one blower compressed the air before it went into the carburetor, and the other blower, or stage of supercharging, further compressed the air after it passed through the carburetor, just before it entered the intake valves.

Any gear-driven supercharger requires power from the engine to turn the impeller at the high speed

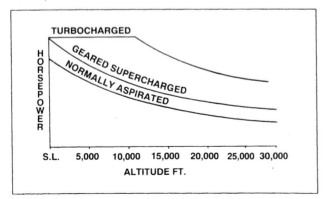

Figure 2-77. *Power vs. altitude.*

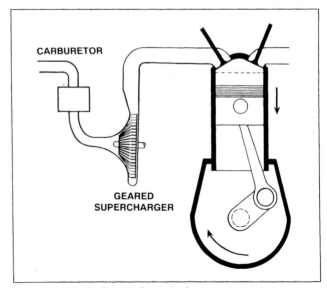

Figure 2-78. *An internal supercharger system.*

required to adequately compress the air, but the loss of power is more than compensated by the power gained by increasing the air density in the induction system.

The most efficient method of increasing the power in modern aircraft engines is by the use of a turbo-supercharger, or turbocharger, as it is commonly called. In this system, a small turbine, attached to a

Figure 2-79. *Turbochargers, driven by the exhaust gases, compress the air before it enters the engine. This allows sea level horsepower to be maintained up to a high altitude.*

centrifugal compressor, is driven by the exhaust gases.

There is, as we know, quite a lot of energy left in the exhaust gases as they leave the cylinders, and some of this energy is extracted to drive the turbine.

The turbocharger, itself consists of an exhaust-driven turbine and a centrifugal compressor on the same shaft. Exhaust gases enter the turbine, spin it, and exit overboard through the exhaust system tail pipe. Air enters the center of the compressor, is compressed, and then flows through the throttle body of the fuel injection system, then through an intercooler and into the cylinders. In figure 2-77, we see that the rated sea level power may be maintained up to considerable altitude without the customary drop-off. This uniform power is maintained by controlling the speed of the turbine.

If all of the exhaust gas flows through the turbine, it will spin at its greatest speed and produce the maximum manifold pressure, but the manifold pressure may be controlled by varying the amount of gas allowed to flow through the turbine. In figure 2-80, we have a complete turbocharger control system. A hydraulically actuated waste gate, spring-loaded to

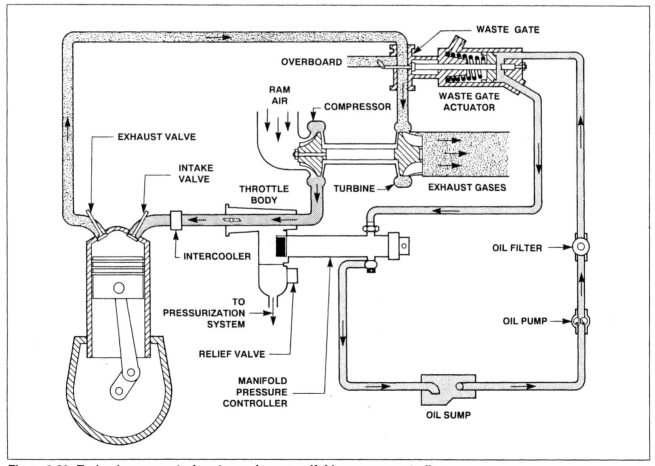

Figure 2-80. *Turbocharger control system using a manifold pressure controller.*

the open position, allows the exhaust gases to dump overboard without passing through the turbine. A manifold pressure controller senses the absolute pressure in the intake manifold, and if it is less than the pressure for which the controller is set, usually around 39 to 40 inches of mercury, it restricts the flow of control oil so the pressure will build up in the waste gate actuator and close the gate against the force of the spring. With the waste gate closed, the exhaust gases flow through the turbine, spinning it, and compressing the air entering the engine. When the desired manifold pressure is reached, the valve in the manifold pressure controller opens, allowing some of

the oil to flow back to the inlet side of the pump. This stops the buildup of pressure in the actuator and holds the waste gate in the proper position to maintain a constant manifold pressure. A relief valve may be incorporated in the system that will automatically relieve any manifold pressure above that set by the engine manufacturer as the maximum allowed. Some of the turbocharger compressed air may be bled off at the throttle body to provide air for the airframe pressurization system.

Modern airplanes are being built with simplicity of operation as the main aim of their manufacturers, and one move toward simplicity is the turbocharger control

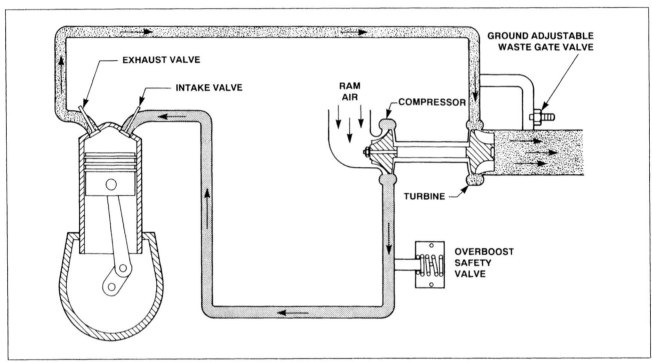

Figure 2-81. Turbocharger control system using a relief valve.

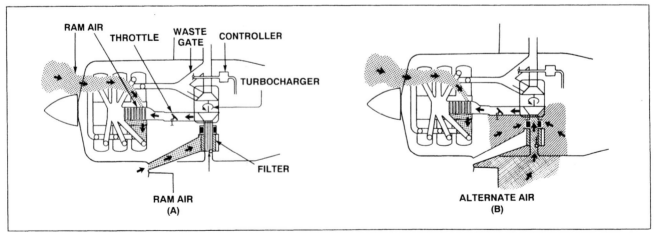

Figure 2-82. The turbocharger can take in its induction air either directly from outside the airplane or through an alternate source where there is less likelihood of ice from flying through freezing rain.

system that operates without a waste gate. In figure 2-81, we see the system. A ground adjustable waste gate valve limits the amount of exhaust gas allowed to flow through the turbine by fixing the size of the bypass around the turbine. The amount of bypass is adjusted so the engine will develop a given manifold pressure at full throttle, at a specified density altitude. A relief valve is installed in the turbocharger discharge and is set above the normal operating pressure to limit the amount of manifold pressure the engine can develop and prevent overboosting.

J. Exhaust System

Volumetric efficiency of an aircraft engine is determined not only by the induction system that directs the air into the engine, but also by the exhaust system that conducts the spent gases out of the cylinders.

The ultimate in simplicity of an exhaust system would be to allow the exhaust gases to leave the engine at the exhaust valve port itself, and this was done in some of the very early engines; but the exhaust valves were subjected to rapid cooling which caused their stems to warp.

The exhaust system evolved through the stages of a collector into which the discharge from several cylinders were piped together and carried overboard, to the present-day system in which all of the exhaust gases are collected and passed through a muffler to reduce the noise level, and then they are discharged overboard.

A typical exhaust collector is seen in figure 2-83. It is constructed of corrosion-resistant alloy steel and bolts onto the exhaust ports of the cylinders.

This collector discharges the gases into a muffler where some of that portion of the energy which causes the sound is dissipated, and then the gases are carried overboard through the exhaust tail pipe. Two examples of aircraft mufflers are shown in figure 2-84. Both of these mufflers are covered with a sheet metal shroud, and air passing between the shroud and the muffler is heated. This heated air may be carried into either the cabin or the carburetor. The lower muffler on the right has a series of short steel knobs welded on its surface to increase its area so

the air can remove more heat for use in the cabin or the carburetor.

When designing an exhaust system, the engineer must consider not only the effectiveness of the muffler and the exhaust system in conducting the gases out of the engine, but he must also consider the amount of back pressure the system creates, and this must be kept to a minimum.

Turbochargers are installed in the exhaust system, and they definitely do cause back pressure; but the design of the turbine is such that it is able to extract the maximum amount of energy with the minimum opposition to the exhaust gases.

Some of the large radial engines have used another method of regaining some of the energy lost in the exhaust gases. Power recovery turbines, called PRT's, were small turbines located in the exhaust system and driven by the exhaust gases. Instead of

Figure 2-84. Exhaust mufflers decrease the noise of the power impulses and provide heat for the cabin and the carburetor.

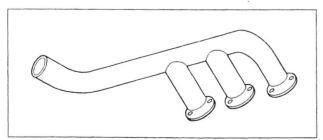

Figure 2-83. Exhaust collector.

driving an air compressor as the turbocharger does, the output shaft of the PRT is geared to the crankshaft through a fluid coupling. This allows some of the energy to be returned to the crankshaft. The weight and complexity of this system made power recovery turbines useful only on some of the largest engines.

K. Ignition Systems

An aircraft engine must draw the proper amount of fuel-air mixture into the cylinder, and this combustible mixture must be ignited when the piston is in the proper position, moving upward on the compression stroke. Ignition usually occurs somewhere around 30° of crankshaft rotation before the piston reaches top center, so the mixture will burn evenly and its maximum pressure will be reached just as the piston passes top center and starts down. Timing the ignition to occur at this point will cause the maximum push to be exerted on the piston.

All certificated aircraft engines have dual ignition, which means that there are two independent ignition systems, firing two spark plugs in each cylinder. The reason for having dual ignition is not only for safety in case one system should fail, but to provide for more uniform combustion within the cylinder. When the spark plugs ignite the mixture, the flame front moves evenly across the piston head from both sides, heating and compressing the mixture in front of the flame. If only one spark plug ignites the mixture, the flame front must move all of the way across the piston, heating and compressing the unburned mixture to the point that it will explode rather than burn.

This instantaneous release of energy is called detonation, and it creates so much highly localized heat and pressure that serious damage will be done to the piston and the cylinder.

1. Magnetos

a. Principle Of Operation

Automobile ignition systems use electrical energy from the battery and generator to provide the spark for igniting the fuel-air mixture, but aircraft engines use magnetos in order to be totally independent of any other system in the airplane. A magneto is a small alternating current generator which produces its electrical current by rotating a permanent magnet in a framework of soft iron around which the coil is wound.

In figure 2-86, we have the principle of an aircraft magneto, but a more detailed description of the entire aircraft ignition system is covered in the IAP, Inc. training manual entitled *Aircraft Ignition and Electrical Power Systems*.

A 2-pole permanent magnet is driven from the crankshaft of the engine, and it rotates between soft iron pole pieces. As it rotates, the lines of magnetic flux flow through the frame in one direction, then stop flowing, reverse, and flow through the frame in the opposite direction. Two coils are wound around the frame of the magneto. One coil, the primary, uses a relatively few turns of heavy wire, while the secondary, wound on top of the primary, has thousands of turns of very fine wire. One of the ends of both coils are joined and connected to ground inside the magneto. The ungrounded end of the primary coil

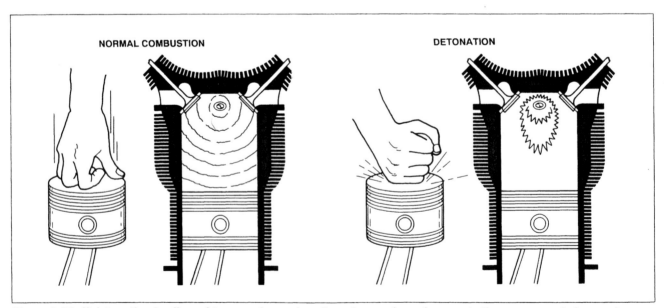

Figure 2-85. If detonation is allowed to occur in an aircraft engine, serious damage to the components is sure to occur.

attaches to one contact of a set of breaker points, and the other contact is grounded. A cam on the same shaft as the rotating magnet opens the breaker points at the proper time.

As the magnet rotates and the magnetic flux in the frame changes, a voltage is induced in the primary winding of the coil which causes current to flow. The breaker points are closed until the flow of primary current reaches the maximum, and then they are opened by the cam. When the breaker points open, the flow of primary current stops; the magnetic field surrounding the coil collapses *immediately*, and this instantaneous collapse induces a very high voltage in the secondary winding.

A capacitor is installed in parallel with the breaker points to prevent arcing as they start to open. Since the primary current is alternating in nature, the capacitor will assist the flow reversal

each time the points close and increase the intensity of the spark.

The secondary winding of the magneto coil goes to the rotor of the distributor which directs the current from the high voltage to the proper spark plug. In the spark plug, the current finds a gap it must jump to ground, and in jumping this gap, the spark is produced which ignites the fuel-air mixture in the cylinder.

b. Timing

(1) Internal

Before a magneto is ready to install on an engine, it must be timed internally so the breaker points will open at the exact instant the flow of primary current is maximum. This is done by adjusting the breaker points so they just break, or open, when the magnet is in what is known as its E-gap position. E-gap is the position of a rotating magnet when it has rotated a specified number of degrees beyond its neutral position. When the magnet is in this position, stopping the flow of primary current will cause the greatest *change* in magnetic flux, and this will induce the highest voltage in the secondary winding.

Special marked teeth on the magnet drive gear and the distributor gear are meshed so the breaker points will open when the distributor finger is in the proper position to route the high voltage to the spark plug.

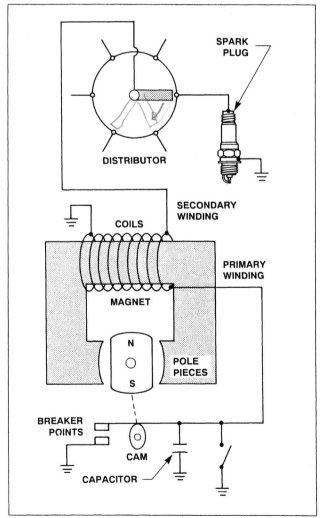

Figure 2-86. *The high voltage for ignition is provided by a magneto which is entirely independent of the aircraft electrical system.*

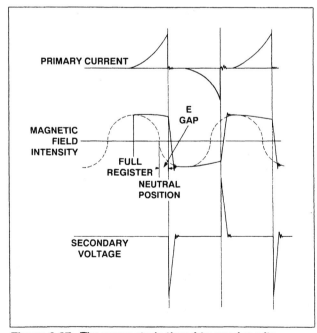

Figure 2-87. *The magneto is timed to produce its greatest voltage at the instant the spark is required at the spark plug.*

(2) Timing the Magneto to the Engine

When the magneto is properly timed internally, with the points just starting to open with the distributor finger positioned for number one cylinder, the magneto is locked, or held in this position. The engine is rotated until number one piston is moving toward top center on the compression stroke and is the proper number of degrees from top center for ignition to occur. The magneto drive gear is then meshed with the accessory drive gear of the engine, and the magneto is secured to the engine with the proper washers and nuts.

To check to be sure the magneto is properly timed to the engine, a timing light is connected across the breaker points of the magneto, and the engine crankshaft rotated backward enough to close the points; then it is turned in the direction of normal rotation. When the crankshaft is in the proper position for firing, the timing light should show that the points are just opening.

2. Spark Plugs

One of the most important yet least appreciated components in an ignition system is the spark plug. The function of a spark plug is to conduct the high voltage from the magneto to an insulated electrode inside the combustion chamber, where it can jump to ground across an accurately controlled gap. This creates the spark that ignites the fuel-air mixture. Spark plugs must operate in an extremely hostile environment, with the tip of the ceramic insulator exposed to temperatures as high as 1,300°F. The tetraethyl lead additive in aviation gasoline forms lead oxides that tend to short out the spark plug, causing it to misfire, and oil and carbon deposits in the firing end of the spark plug prevent its accomplishing its purpose.

Almost all modern aircraft spark plugs are shielded; that is, they are completely encased in a steel shell, and the high voltage is carried by a special harness to the center electrode, mounted in the special, glass-hard ceramic insulator, figure 2-88.

The center electrode is made of a nickel alloy, filled with copper so it can more effectively conduct the heat into the insulator. Two, three, and sometimes four ground electrodes are bonded into the opening of the firing end cavity to provide a surface to which the spark can jump. The distance between the surface of the ground electrodes, and the center electrode is critical and must be reset at each spark plug service interval to compensate for the material that is eroded away each time a spark jumps.

Most spark plugs have a resistor built into the center electrode to minimize the amount of current that is allowed to flow for each spark. Electrical energy is stored by the capacitive effect of the harness shielding, and if there were no resistor in the spark plug, this energy would cause a spark of excessively long duration and would unduly erode the electrodes.

When lead fouling problems are excessively severe in an engine, a fine-wire spark plug such as the one in figure 2-89 may be used. Fine-wire spark plugs have a much more open firing end to aid in scavenging the lead-rich combustion gases, and the fine-wire electrodes made of platinum or iridium can operate at temperatures high enough to prevent lead oxides forming inside the cavity.

3. Ignition Leads

The high voltage is generated in the magnetos and carried to the spark plugs by a special high-voltage ignition lead. This lead is usually made of stranded,

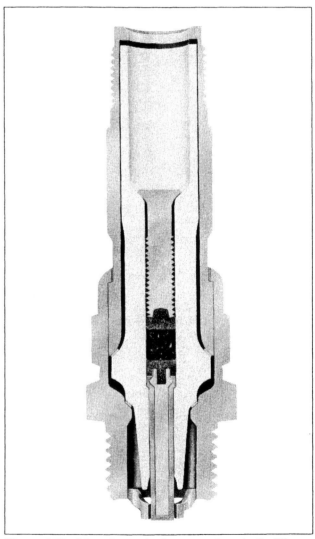

Figure 2-88. Aircraft spark plugs are almost all of the shielded type in which the radiated energy is passed to ground rather than being allowed to cause radio interference.

stainless steel wire and attaches to the magneto through a removable plate which holds all of the leads tightly into the distributor block. Contact is made with the center electrode of the spark plug with a coil spring held in a silicone-rubber insulator. The conductor is covered with a special high-voltage insulation and this is enclosed in a braided wire shield, grounded at both of its ends, to conduct to ground the electrical energy radiated from the conductor when the spark occurs, preventing radio interference. The braid is enclosed in a plastic covering to protect it from abrasion.

L. Propeller

While the propeller is actually not a part of the engine itself, it is an essential part of the propulsion system,

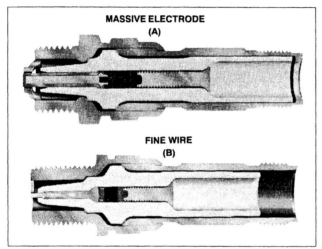

Figure 2-89. *Fine wire spark plugs may be used if lead fouling conditions are so severe that the massive electrode spark plugs become lead fouled.*

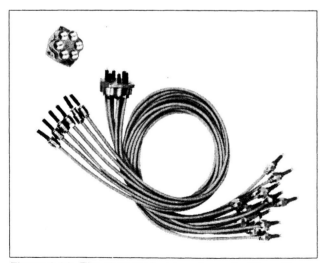

Figure 2-90. *The electrical energy is carried from the magneto to the spark plug through specially shielded high voltage ignition leads.*

as it is needed to convert the rotary motion of the crankshaft into thrust to move the airplane through the air.

A propeller is essentially a rotating airfoil that produces thrust in the same way a wing produces lift. A review of basic aerodynamics tells us that the lift produced by an airfoil is determined by the shape of the airfoil section, the angle that the airfoil strikes the air, the speed it moves through the air, and the air density. The air density is the same all along the propeller, but the speed of the blade relative to the air varies from a minimum at the root to a maximum at the tip; and in order to produce a thrust force that is rather uniform along the length of blade, both the cross section and the angle change along the blade. The root is shaped to produce its thrust at a relatively slow speed, and it has a large pitch angle. At the mid-portion of the blade, the section is much thinner, and has a lower blade angle, requiring more speed for its thrust; at the tip, the angle is the lowest and the section is the thinnest. Varying the shape and the angle of the blade as its speed varies makes it possible for the designers to distribute the forces along the blade and get the most thrust for the power developed by the engine.

For an engine to develop its maximum power, as it must do for takeoff, it must turn at a high speed, and to do this, the propeller must have a low pitch angle. But when the airplane is in the air, in order to get the highest cruising airspeed for an economical engine RPM, the propeller must have a high pitch. The most simple airplanes use a fixed-pitch propeller, and a compromise must be made between takeoff performance and cruise speed.

More efficient performance of an airplane may be had if the propeller is capable of having a flat pitch for takeoff, and after the airplane is airborne and the

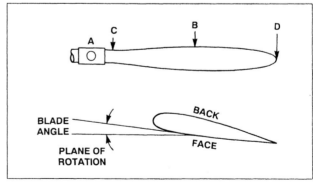

Figure 2-91. *Propeller nomenclature.*
- (A) *Hub.*
- (B) *Blade.*
- (C) *Bladeroot.*
- (D) *Blade tip.*

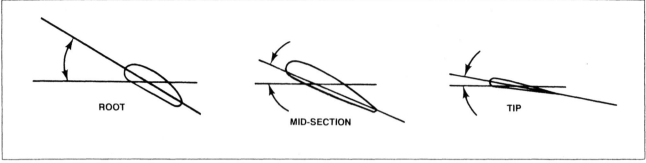

Figure 2-92. *The blade angle of a propeller decreases along the length of the blade so that the increased velocity near the tip will not produce undue strain on the blade.*

power is reduced, the pitch can be increased. This may be done with a 2-position propeller.

When the propeller is spinning, centrifugal force acting on the blade will tend to rotate it around its pitch change axis and move the blades into a low pitch. This force is called centrifugal twisting moment (CTM). A hydraulic cylinder is built into the propeller hub and the piston acts through a rod attached to the blade root to put the blade into high pitch. Some propeller designs make use of higher forces by building counterweights into the blade roots that apply a strong force on the blade, rotating it into a high pitch. When counterweights are used, the piston rod is attached to the blade root in such a way that it will overcome the counterweight force and work with the CTM to move the blades toward low pitch.

Even more efficient operation may be had if the oil being directed into the pitch change mechanism of the propeller is controlled by a flyweight-type governor that will either direct oil into the propeller or

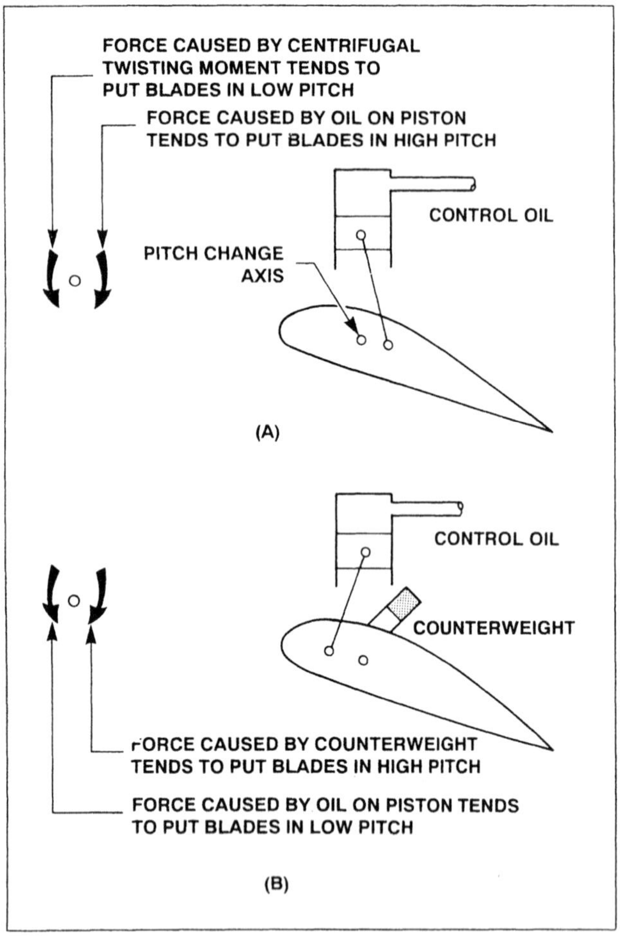

Figure 2-93.

   (A) *Non-counterweight propellers use oil pressure to move the blades toward high pitch and centrifugal twisting moment to place them in low pitch.*

   (B) *Counterweight propellers use the centrifugal force acting on the counterweights to move the blades into high pitch and oil pressure to move them into low pitch.*

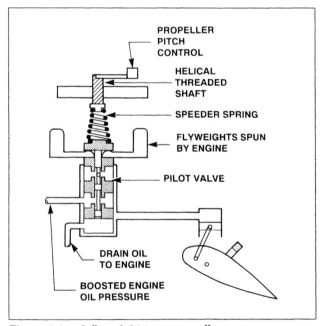

Figure 2-94. *A flyweight-type propeller governor maintains the propeller RPM as set by the pilot by varying the propeller pitch as the propeller airloads vary.*

drain it out to hold the engine RPM constant at the speed the pilot sets. In figure 2-94, we have a schematic of the operation of a governor controlling a counterweight propeller. The propeller pitch control adjusts the compression of the speeder spring which opposes the centrifugal force acting on the spinning flyweights. When the engine is not up to the speed set by the pitch control, the speeder spring holds the pilot valve all of the way down, and the engine oil pressure boosted by a small pump inside the governor forces the blades into low pitch. When the engine speed builds up to that called for by the pilot, the centrifugal force on the flyweights overcomes the compression of the speeder spring and lifts the pilot valve until it closes off the line to the propeller. The pitch is held constant since the piston will not allow the force on the counterweights to increase the pitch. Now, if the nose of the airplane pitches down slightly, the propeller load will decrease, and the engine will attempt to speed up. The force on the flyweights will increase enough to overcome the compression on the speeder spring, raising the pilot valve and allowing some of the oil from the propeller to drain back into the engine so the force on the counterweight can increase the pitch of the propeller and bring the engine speed back to that set by the governor. Keeping the engine speed constant, by adjusting the propeller load, allows the engine to operate most efficiently.

Study Questions

1. Why does operating an aircraft engine for only an hour or so a week tend to decrease its allowable time between overhauls?

2. What is the main disadvantage of an in-line engine for aircraft use?

3. What was one of the advantages of the V-type engine that caused it to be used in some of the high-performance fighter aircraft of World War II?

4. What was one of the main advantages of the rotary-radial engine used in World War I over the other engines available at the time?

5. What was the main advantage of the static-radial engine that caused it to become the standard of aircraft engines for so many years?

6. What are two reasons the horizontally opposed engine has become the most popular configuration for aircraft engines for modern general aviation aircraft?

7. What is used as a coolant in high-performance liquid-cooled aircraft engines?

8. What is meant by pressure cooling of an aircraft engine?

9. What is meant by a wet sump engine?

10. Where is number 1 cylinder on a 14-cylinder radial engine?

11. What is the number of the cylinder nearest the propeller on the left side of a V-12 engine?

12. Where is number 1 cylinder on a 6-cylinder Lycoming opposed engine?

13. Where is number 1 cylinder on a 4-cylinder Continental opposed engine?

14. What is the firing order of a 9-cylinder radial engine?

15. What is the firing order of an 18-cylinder radial engine?

16. What is the firing order of a 4-cylinder Lycoming opposed engine?

17. What is the firing order of a 6-cylinder Continental opposed engine?

18. Of what material are the cylinder barrels of an air-cooled engine made?

19. How are the cylinder heads attached to the barrels of most air-cooled aircraft engines?

20. What are two advantages of a chromed cylinder over a plain cylinder?

21. What cylinder barrel treatment is identified by a band of orange paint around its base?

22. What type of cylinder wall treatment is known as nitriding?

23. What is one of the main problems with nitrided cylinder walls?

24. How are valve guides held in place in an aircraft engine cylinder head?

25. Who are some exhaust valves filled partially with metallic sodium?

26. Of what material are most aircraft engine pistons made?

27. What is the purpose of the top piston rings in an aircraft engine?

28. What is the purpose of the piston ring directly below the compression rings?

29. If there is a ring installed near the skirt of the piston, what is its function?

30. What kind of piston ring would be most properly used in a cylinder having chromed walls?

31. How are the piston pins held in the pistons and connecting rods on most modern aircraft engines?

32. How is it possible in a radial engine to attach the connecting rods of nine cylinders to one throw of the crankshaft?

33. How many degrees apart are the throws of the crankshaft of a 6-cylinder horizontally opposed aircraft engine?

34. What treatment is given the bearing journals and pins of a crankshaft to increase its wearing ability?

35. What is the purpose of the dynamic dampers on an aircraft crankshaft?

36. Who is the propeller on some aircraft engines geared to turn more slowly than the crankshaft?

37. Why do some geared aircraft engines drive the propeller shaft drive gear through a quill shaft?

38. What is the gear ratio of a planetary gear arrangement which has 72 teeth on the drive gear and 36 teeth on the fixed gear?

39. Which clearance is greater on a radial engine: the hot, running valve clearance, or the cold clearance?

40. What is the purpose of a hydraulic valve lifter in a horizontally opposed engine?

41. Of what material are most aircraft engine crankcases made?

42. What type of oil seal is used to seal a rotating shaft as it comes out of a crankcase?

43. What is one of the main disadvantages of straight mineral oil for use in aircraft engines?

44. Why is detergent oil unsatisfactory as a lubricant in aircraft engines?

45. What type of additives are used in AD-type aircraft engine oil?

46. Is it harmful for engine operation to mix different brands of AD oil in the engine?

47. What is the SAE number of Aviation 80 oil?

48. How does the oil get to the valve operating mechanism on the cylinder heads to lubricate it?

49. What prevents cold oil being further cooled in the oil cooler?

50. Is a gerotor-type of oil pump a constant or a variable displacement pump?

51. What is the purpose of a compensated oil pressure relief valve?

52. What is the purpose of an oil dilution system in an aircraft engine?

53. Which type of filter, bypass or full-flow, is used in most modern aircraft engines?

54. What is one of the disadvantages of a depth-type oil filter?

55. How is the element in a sealed-type oil filter examined for metal contamination when the filter is removed during an inspection?

56. How is fluctuation of the oil pressure gage minimized?

57. Is oil temperature taken as the oil enters the engine, or as it leaves?

58. What is the purpose of the thin, sheet metal baffles around the fins of an air-cooled engine cylinder?

59. How do cowl flaps aid in cooling a horizontally opposed engine?

60. How does an augmenter tube increase the cooling of an aircraft engine?

61. What are four functions of a fuel metering system for an aircraft engine?

62. What forces the fuel out into the airstream of a float-type carburetor?

63. In a float-type carburetor, what determines the maximum amount of fuel that can flow for any given pressure drop?

64. How is the fuel metering force generated in a typical pressure carburetor?

65. Does the fuel-air mixture tend to enrich or lean as an airplane goes up in altitude?

66. What is changed to vary the idle RPM of an aircraft engine?

67. What is the function of an acceleration system on an aircraft fuel metering system?

68. How does a power enrichment system prevent an aircraft engine overheating at full power operation?

69. What is the main difference between a carburetor and a fuel injection system?

70. Who is the fuel-air mixture distribution more uniform with a fuel injection system than it is with a carburetor?

71. Name three types of air filters used in modern aircraft engines.

72. How are paper air filters cleaned?

73. What is the danger of using carburetor heat when the engine is operating at full throttle?

74. What is one of the main reasons for limiting the use of carburetor heat when operating the engine on the ground?

75. What is the highest manifold pressure it is possible for a normally aspirated aircraft engine to produce?

76. What is meant by a ground-boosted engine?

77. What drives the compressor of a turbocharger?

78. How is the speed of a turbocharger controlled?

79. What is sensed by a turbocharger controller to vary the position of the waste gate?

80. Of what material are most aircraft engine exhaust systems made?

81. What two uses are made of the air that is heated by passing it over parts of the exhaust system?

82. Why is it important that aircraft engine exhaust systems create the least back pressure possible?

83. How does a power recovery turbine utilize part of the energy left in the exhaust gases?

84. Where is the piston when ignition normally occurs in an aircraft engine?

85. What are two reasons for using a dual ignition system in an aircraft engine?

86. Why are magnetos preferred for aircraft engine ignition rather than a battery ignition system such as that used in automotive engines?

87. Essentially, what is a magneto?

88. Is the spark produced in a magneto when the breaker points open or when they close?

89. What are two purposes of the capacitor in an aircraft magneto?

90. What relationship must be established when internally timing an aircraft magneto?

91. What is meant by E-gap?

92. What is the function of a spark plug in an aircraft engine?

93. What is an advantage of a fine-wire spark plug over a massive-electrode spark plug?

94. Why do some spark plugs have a resistor built into their center electrode?

95. What is the purpose of the wire braid enclosing the spark plug leads?

96. What is the basic purpose of a propeller?

97. Should a propeller have a low- or a high-pitch angle for the engine to develop its maximum power?

98. What two forces are used to vary the pitch on a constant speed propeller?

Chapter III

Engine Service And Overhaul

A. Determination of Service Life

The design service life of the World War I Curtiss OX-5 engine was only fifty hours, and while this seems absurdly short to us today, we must remember that in its fifty hours of life, it had probably outlived several airplanes and more than likely, more than one student or instructor.

When the economics of civilian aviation began to dictate the service life of engines, the time increased. Better materials, better designs, and knowledge of proper operation increased the time to 100, 600, and then 1,000 hours, and after we entered the space age, new and exotic valve materials allowed the service life to go up to as much as 2,000 hours between overhauls.

Today, when a decision is made on which airplane to buy, the manufacturer's recommended TBO (time between overhauls) is one of the factors to be considered.

1. Manufacturer's Recommended TBO

With the cost of overhauls continually increasing, economical operation of an airplane requires that an engine run a maximum number of hours before it requires a complete disassembly and overhaul. The hours given in the manufacturer's TBO are recommended hours and are no guarantee that the engine will actually run that long, and they most certainly do not even imply that it will operate this long without maintenance. On the other hand, there is no implication that when an engine reaches the recommended TBO that it must be given a major overhaul. Some operators however, specify in their operations manuals that the engine be overhauled at the manufacturer's TBO; but for the private owner, there is no such requirement.

What is actually meant by the manufacturer's recommended TBO is that if the engine is operated under average conditions with the maintenance done according to the manufacturer's recommendations, there should not be enough wear to render the reusable parts beyond their serviceable limits. This allows major overhaul to be done economically.

When *average* operating conditions are specified, this naturally excludes such things as operating in excessively dusty or sandy conditions, operating at continually high power, or exceeding the allowable RPM. Defective cooling baffles, worn air filters, improper

grade of fuel, inadequate oil change intervals are only a few of the operational problems that can immeasurably shorten the time between overhauls.

The overhaul specified in the manufacturer's TBO is a major overhaul and does not include a top overhaul. Worn piston rings, burned valves, or lead accumulation on the valve stems may re quire that the cylinders be removed and over hauled. This constitutes a top overhaul, and because of the many variables that can damage the cylinder assemblies, there is no specified time for top overhauls.

Many operators consider a top overhaul poor economy, but there are times when it can be wise to perform a top. If an engine has run to almost its recommended TBO, it would be uneconomical to perform a top overhaul and then just a few hours later be required to repeat the operation in a major overhaul. But if there was some obvious cause for cylinder damage, such as broken baffles, leaking carburetor air filter, clogged fuel injection nozzles, or any other localized condition that damaged the cylinders while there were relatively few total hours on the engine, it would be expedient to overhaul the cylinders and extend the life of the engine.

If an engine has had good maintenance, has been held carefully within the operating limits allowed by the manufacturer, when it reaches the recommended TBO, it may be given a careful inspection, especially noting the differential compression and the oil consumption, and a spectrometric inspection should be made of a sample of the engine oil. If everything appears to be good to the A&P, he can certify the engine as airworthy and release it for another one hundred hours of operation. A continual watch is kept of the compression and oil consumption, and a careful observation is made of the contamination growth trend as shown in subsequent spectrometric oil analysis.

When an engine is operating beyond its TBO, it is a good idea to consult with the engine manufacturer and follow his recommendations to know exactly how long this extension should be allowed. The danger of catastrophic failure of the engine itself is minimal if these precautions are followed, but operating vital engine accessories beyond their normal overhaul time may be courting disaster. Check with the engine manufacturer and wisely follow his experience-based recommendations.

2. Attrition Factors

a. Wear

The economy of an aircraft engine is based not only on the number of hours allowed between overhauls, but the number of parts that must be replaced when the engine is overhauled.

Operating too long between oil changes or with inadequate filtering will allow the oil to carry particles through the engine which act as a fine abrasive, wearing out the engine parts. Leaking air filter boxes or damaged air filters can allow sharp sand crystals to enter the engine and, in a short time, wear the cylinder walls and piston rings. The excessive use of carburetor heat on the ground will also allow unfiltered air into the engine.

Abrupt power changes can cause the cylinders to shrink enough around the heat expanded piston to cause piston scuffing and shorten the life of the engine.

Operating an engine with fuel having more tetraethyl lead than the engine is designed to assimilate will lead to spark plug fouling, oil contamination, and sticking valves.

The increased cost of aviation gasoline has made many pilots more conscious of the mixture control than they have been in the past, and this can be a blessing as well as a curse for the engine life. Operating with too rich a mixture will cost more, not only in terms of the fuel used, but will lead to oil contamination and spark plug fouling, both of which require additional maintenance. Excessive leaning of the mixture can cause damage because of the higher temperature of the departing gases, and so one of the most useful instruments a pilot has at his disposal is the exhaust gas temperature gage, which, if used properly, can allow the pilot to operate his engine with the most economical fuel-air mixture without the danger of burned valves or preignition. More will be said about this instrument in the section on troubleshooting.

If an engine has been allowed to operate beyond its red-line RPM, or if its cylinder head temperature has been in excess of that allowed by the manufacturer, there is a good possibility that wear of the engine parts will be excessive. Even if the engine should be able to reach its TBO, the number of parts requiring replacement will be greater than that for a normal overhaul.

b. Corrosion

Wear affects an aircraft engine during its operating life as the moving parts rub against each other, but actually more damage is done to an aircraft engine when it is not operating than when it is used every day. The static attritional force is corrosion.

Corrosion is an electrochemical process that forms inside an engine where dissimilar metals are in contact with each other in the presence of an electrolyte. When the engine is not being operated, moisture will condense inside the crankcase and react with the aluminum, bronze, steel, and other metals to form corrosion. When an engine is run for only short periods of time, the oil does not have a chance to thoroughly heat up, and moisture will collect in the oil and react with the sulfur to form sulfuric acid. This is extremely potent as an electrolyte and will cause extensive corrosion damage inside the engine. Many cylinders and crankshafts have their surfaces case hardened by a process known as nitriding by which the surface of the metal is converted into an extremely hard material. This nitrided surface is highly resistant to wear but is susceptible to pitting from moisture.

Any time an aircraft engine is allowed to sit for a period of time out of service, it should be protected from corrosion by spraying the inside of the cylinders with a tenacious rust-preventive oil which will cling to the cylinder walls. After the oil is sprayed in, the engine should not be turned, as the movement of the pistons will wipe away the protective coating.

c. Improper Maintenance

The service life of any aircraft engine may be increased by proper maintenance. Small discrepancies such as loose exhaust nuts can, if they are not attended to, cause a blown gasket and a damaged mounting surface. Such a simple act as tightening a nut will, if not done in time, require the replacement of a cylinder. Cracks in an exhaust system will grow, and loose induction system packings will leak and cause overheating and detonation.

Careful inspections by knowledgeable A&Ps with immediate correction of any discrepancies found will do more to increase the time between overhaul of the engine than any other one thing.

B. Classifications of Servicing

1. Inspections

There is possibly no aspect of engine servicing as important as the routine inspections. When each of these is diligently carried out, impending problems may be found before they can become major, and in addition to increasing the safety of flight, aircraft operation will be made more efficient.

a. Preflight Inspection

The Federal Aviation Regulations require that before any pilot begins a flight, they satisfy themselves that, among other things, the airplane is in all regards safe enough for the flight. This can only be done by a careful preflight inspection, and here we will be concerned with only that part of the inspection that applies to the powerplant.

Before starting an actual preflight inspection, it must be determined that the ignition switch is in the off position; then if the cowling can be opened, a visual inspection should be performed on the engine.

- Starting at the rear of the accessory section near the firewall, inspect all of the wiring and plumbing to be sure none of it is loose, chafing against any of the engine components, or showing any signs of wear or deterioration.
- Check the oil to be sure that it is within the operating level specified by the aircraft flight manual. (Most engines have a tendency to throw out the top quart of oil, so for most local flights, they are operated less than full.)
- Check all of the wire connections to the magneto to be sure that there are no loose connections or chafed wiring.
- Check all of the wiring to the generator or alternator, and to the voltage regulator. If the battery and master relay are ahead of the firewall, check them as well. Gently shake all of the wires to be sure that none of them are loose at the connections.
- Check the accessory section of the engine compartment for indications of fuel or oil leaks. Check all of the baffles for integrity of the air seals and those around the cylinders for cracks or broken sheet metal that could cause local hot-spots on the cylinders.
- Check the paint on the cylinders for any indication of discoloration which could indicate that detonation has taken place inside the cylinders.
- Check the primer lines as they enter the cylinders or, if the engine is fuel injected, the injector lines where they attach to the nozzles.
- Check the push-rod tubes and rocker box covers for indications of oil leakage.
- Check the intake pipes for any traces of dye from the fuel which might indicate an induction system leak.
- Check the spark plug leads for their general condition and for any looseness where they screw into the spark plugs. Be sure that all of the leads are secured in such a way that they cannot be burned by the exhaust stacks.
- Carefully examine the exhaust connections where the pipes join the cylinders to be sure that there are no blown gaskets and all of the nuts are in place.

- Drain the main fuel strainer and examine the gasoline that came out of it. There should be no indication of water in the fuel.
- Inspect the carburetor air filter for both its security of mounting and cleanness
- Check as much of the induction system as is visible for traces of fuel dye stain.
- Check the propeller for nicks or scratches, and all of the attachment bolts or nuts for safety.

When the engine is run up, it should develop the required static RPM and manifold pressure, and the magneto drop should be within the limits specified by the manufacture with equal drops on both magnetos. A magneto switch check should be performed to be sure the switch is operating properly, and the oil and fuel pressure should be within the proper operating range. The propeller should cycle smoothly between low and high pitch, and there should be the proper drop in RPM when the carburetor heat control is pulled.

b. 50-Hour Inspection

Though this inspection is not required by the FAA, the engine manufacturers have proven that by observing it, the life of the engine may be in creased. A typical 50-hour inspection consists, in addition to all of the items performed on the preflight inspection, of removing the cowling and examining the:

(1) Ignition System

- The spark plug leads should be checked for security and for any indication of corrosion.
- All of the leads should be tight in both the spark plug and the magneto distributor block, and there should be no chafing or wear.
- The spark plugs should be examined where they screw into the cylinder head for any indication of leakage of the hot gases from the cylinder.

(2) Fuel and Induction System

- Check the primer lines for indication of leaks, and for security of the clamps. Remove and clean the fuel inlet strainers.
- Check the mixture control and throttle linkage for travel, freedom of movement, security of the clamps, and lubricate the controls if necessary.
- Check the air intake ducts for leaks, for security, for filter damage or evidence of dust or other solid contaminants that may have leaked past the filter.
- Check the fuel pump vent lines for any evidence of fuel or oil seepages which could indicate the failure of one of the seals.

(3) Lubrication System

• Drain and replace the engine oil if recommended by the engine manufacturer. Under some conditions when a full-flow filter is installed, the oil drain period may be increased, but in almost all circumstances, the filter should be replaced every 50 hours.

• Remove the full-flow filter if one is installed, and cut the element open to inspect it for any traces of metal particles which would most likely indicate an impending engine failure, and check all oil lines for any indication of leakage, or for signs of chafing.

(4) Exhaust System

• Check all of the flanges in the exhaust pipes where they attach to the cylinder heads for evidence of leakage. If they are loose or show any signs of leakage, they must be removed and machined flat before they are reassembled.

• Check the entire exhaust manifold and muffler for their general condition.

(5) Cooling System

• Check the cowling and baffles for any indication of damage or missing parts, and the entire system for security.

(6) Cylinders

• Check the rocker box covers for indication of leaks, and replace the gasket if leaks are found. Carefully check the entire cylinder for indication of overheating which would cause the paint to be burned or discolored. This would indicate detonation and would make further inspection by borescope or cylinder removal necessary.

• Check for any indication of discoloration of seepage between the cylinder head and barrel.

• Check between the fins of the cylinder head for any indication of cracks.

(7) Turbocharger

• Check all of the oil lines for leaks or chafing, and check all brackets for security and for any indication of damage or cracks.

• Check the waste gate for freedom of action and the alternate air door for operation and sealing.

c. 100-hour or Annual Inspections

Perhaps the most important maintenance tool for prolonged engine life is the required 100-hour or annual inspection. The actual inspections are identical, the only difference being the person authorized to perform them. A 100-hour inspection is required for every aircraft operating for hire under the inspection program specified in FAR 91.169, and the inspection may be performed by an A&P technician. All aircraft operating under this part of the FAR, whether they are being operated for hire or not, must have an annual inspection which must be performed by an A&P holding an Inspection Authorization.

(1) Preliminary Paperwork

One of the most important aspects of this inspection is the paperwork involved. Before starting the actual inspection, check all of the aircraft records:

• Check and list all of the Airworthiness Directives against the engine and all of its components, such as carburetors, magnetos, alternators or generators, propellers, and ignition switches.

• Check the General Aviation Inspection Aids to see the problems that have been found with similar engines.

• Review the manufacturer's service bulletins and service letters to be sure that there is nothing that should be done to the engine to make it safer or more efficient.

A shop work order is started for the inspection and the records are made ready to be examined.

• Be sure that all of the accessories on the engine are of the approved type and everything that is included in the equipment list as being installed is actually on the engine.

• Be sure the propeller is the proper model, and its blades are approved and are the ones listed as being installed.

• Check the total time on the engine and propeller, and compare this with the list of life-limited parts to see if any of them are nearing their retirement time.

• Check to see if there are any Major Repair and Alteration Forms (VA 337) on the engine or propeller, and if there are, the repair or alteration must be checked to see that it is actually as described on the form.

(2) Pre-inspection Run-up

After all of the paperwork has been inspected, the engine is given a good pre-inspection run-up to determine its actual condition and get the oil warm and the cylinder walls well oiled.

• Check all of the temperatures and pressures to be sure they are in the proper operating range.

• Check to see that the engine will develop its proper static RPM and that the magneto drop is within the range recommended by the aircraft manufacturer, and that the drop is uniform on both magnetos.

• Check for any abnormal noises in the engine, and for any vibrations that are not characteristic of that engine.

ANNUAL INSPECTION FORM

Make & Model _____ N_____ Date_____

Serial No._____ Year_____ Tach Time_____

Owner _____ Airframe Total _____

Address _____ Engine Total_____

_____ Engine TSMOH _____

ITEM	MAKE	MODEL/PART NO.	SERIAL NO.	APPROVED
Engine				
Prop				
Carb				
Mags: Left				
Right				
Generator/Alternator				
Starter				
Misc. Acces.				
ELT				
Altimeter				
Seat Belts				

II

AD NO.	DESCRIPTION	COMPLIANCE	ADDITIONAL COMPLIANCE

Figure 3-1. Typical annual inspection form (1 of 6).

III   Flight Manual required: Yes_____No_____

Special Markings/Placards required:_____

_____

_____

_____

_____

_____

_____

IV Aircraft papers in order:	Yes	No
Registration		
Airworthiness Certificate		
Radio License		
Flight Manual		
Special Markings		

V   Altimeter Static Test [FAR 91.170] Date of Compliance _____

VI   Special Inspections [Service Manuals, Bulletins, Inspec. Aids, etc.]

Reference	Inspection	Completed

Figure 3-1. Typical annual inspection form (2 of 6).

INSPECTION REPORT

This form meets requirements of FAR Part 43   Work Order No._____

Make	Model	Serial No.	Registration No.
Owner		Date	
Type of Inspection		Tach Time	

		L	R	100	500	insp.

A. PROPELLER GROUP

1. Inspect spinner and back plate...
2. Inspect blades for nicks and cracks ....
3. Inspect hub for cracks and corrosion ....
4. Check for grease and oil leaks ....
5. Check mounting bolts and safety...
6. Constant speed — check blades for tightness in hub pitot tube ....
* 7. Constant speed — remove prop, remove sludge...
* 8. Lubricate as per manual...
9. Inspect complete assembly ....
10. Replace spinner ....

B. ENGINE GROUP

1. Remove engine cowls
2. Clean cowling, check for cracks, missing fasteners, etc. ....
3. Compression check: /80...
   L. #1 #2 #3 #4 #5 #6
   R. #1 #2 #3 #4 #5 #6
* 4. Drain oil..
5. Check oil screens and clean ....
6. Replace oil filter element...
7. Check oil temp sender unit for leaks and security ....
8. Clean and check oil radiator fins..
* 9. Remove and flush oil radiator..
10. Check and clean fuel screens..
11. Drain carburetor...
*12. Service fuel injector nozzles ....
13. Check fuel system for leaks...
14. Check oil lines for leaks and security ....
15. Check fuel lines for leaks and security ....
*16. Service air cleaner ....
17. Check induction air and heat ducts ....
18. Check condition of carb heat box ....
19. Check mag points for proper clearance ....
20. Check mags for oil seal leakage..
21. Check breaker felts for lubrication ....
22. Check distributor block for cracks, burned areas,
    corrosion, height of contact springs...
23. Check ignition harness and insulators ....

Figure 3-1. Typical annual inspection form (3 of 6).

	L	R	100	500	insp.

24. Check mag to engine timing:

        Left _____ ....

        Right_____ ....

*25. Service or replace spark plugs ....

26. Check condition of generator or alternator ....

27. Check condition of starter ....

28. Check condition and tension of drive belts ....

29. Check hydraulic pump and strainer ....

30. Check vacuum pump and lines ....

31. Inspect exhaust stacks, gaskets, etc. ....

32. Inspect muffler and shrouds ....

33. Check engine baffles ....

34. Check breather tube for obstructions, security ....

35. Check crankcase for leaks, cracks, etc. ....

36. Check engine mounts for cracks, loose mounts ....

37. Check engine mount bushings ....

38. Check firewall seals ....

39. Check throttle, carb heat, mixture, and prop governor

    controls for travel and operating condition ....

40. Check cowl flap condition and operation ....

41. Inspect engine for general condition, loose parts, chafing,

    proper safeties, proper installation ....

*42. Fill engine with oil ....

43. Clean engine ....

44. Lubricate all controls ....

45. Reinstall engine cowl ....

C. CABIN GROUP

1. Inspect cabin doors for damage and operation ....

2. Check windows for general condition ....

3. Check upholstery for general condition ....

4. Check seats, seat belts, and mountings ....

5. Check trim operation ....

6. Check flap operation ....

7. Check fuel selector valve operation ....

8. Check operation of fuel drain ....

9. Check landing, navigation, cabin, and instrument lights ....

10. Check rudder pedals ....

11. Check brake cylinders for operation and leaks ....

12. Check brake fluid level ....

13. Check control wheels, column, pulleys, cables ....

14. Check instruments, lines, and attachments ....

*15. Service filters for gyro instruments ....

16. Check condition in general under panel for loose wires, loose equipment, chafing, etc. ....

17. Check condition of heater controls and ducts, air vents, and air conditioning ducts ....

18. Lubricate controls ....

19. Check condition of instrument panel shock mounts ....

20. Check altimeter calibrtation ....

D. FUSELAGE AND EMPENNAGE GROUP

1. Remove inspection plates and panels ....

2. Check general condition of aircraft skin ....

3. Check baggage door[s] latches, and hinges ....

Figure 3-1. Typical annual inspection form (4 of 6).

	100	500	insp.
* 4. Service battery			
5. Check bulkheads and stringers for damage			
6. Check wiring for damage and security			
7. Check security of all lines [fuel, hydraulic, etc.].......................			
8. Check cables, turnbuckles, guides, and pulleys for safeties, damage, and operation			
9. Check radio antennas for mounting and electrical connections............................			
10. Check rotating beacon or strobes for security and operation........................			
11. Check empennage surfaces for damage			
12. Check rudder hinges, horn, and attachments			
13. Check vertical fin attachments			
14. Check elevator or stabilator hinges, horn, and attachments..........			
15. Check horizontal stabilizer attachments			
16. Check trim mechanism..			
*17. Service hydraulic and brake systems			
*18. Service cabin heater			
*19. Lubricate as per manual..			
20. Reinstall inspection plates and panels			

E. WING GROUP

1. Remove inspection plates and fairings			
2. Check surfaces and tips for damage, loose rivets, etc.			
3. Check ailerons, hinges, cables, pulleys and bellcranks for damage and operation			
4. Check flaps for damage and operation			
5. Check fuel tanks for leaks and water			
6. Check fuel tank vents			
7. Fuel tanks marked for octane and capacity			
8. Check wing attachment bolts			
9. Check pitot tube for security			
10. Check stall warning indicator for operation			
11. Lubricate as per manual..			
12. Reinstall inspection plates and fairings			

F. LANDING GEAR GROUP

1. Put airplane on jacks			
2. Check tires for wear and damage..			
* 3. Remove wheels for cracks, corrosion, broken bolts			
4. Check tire pressure N_____M_____			
5. Check brake linings and drum or disc			
6. Check brake lines for security and condition			
7. Check oleo struts for leaks and scoring			
8. Check gear forks, torque links, etc., for general condition............			
9. Check gear attachment bolts			
10. Check nose gear steering control and travel			
11. Check shimmy dampener..			
12. Check gear doors and attachments			
13. Retract gear, check for proper clearances and operation................			
14. Check warning horn and lights			
15. Check safety "squat" switch..			
16. Check oleo fluid level and air pressure			
17. Inspect all hydraulic lines and electrical wires for security, routing, and general condition................			
*18. Lubricate as per manual..			
19. Remove airplane from jacks			

Figure 3-1. Typical annual Inspection Form (5 of 6).

G. OPERATIONAL INSPECTION

	L	R	100	500	insp.
1. Check fuel pump					
2. Check fuel quantity and pressure- or flow-gage					
3. Check alternator or generator output					
4. Check vacuum gage					
5. Check gyros for noise and roughness					
6. Check cabin heater operation					
7. Check electronic equipment operation					
8. Check air conditioner operation					
9. Check parking brake					
10. Check oil pressure and temperature					
11. Check manifold pressure					
12. Check alternate air					
13. Check propeller smoothness					
14. Check prop governor action					
15. Check mag RPM variation L. Engine R. Engine					

    L.
Mag
    R.

	L	R	100	500	insp.
16. Check static RPM _____					
17. Check idle RPM _____					
18. Check mag switch operation					
19. Check throttle and mixture operation					
20. Check idle mixture					

H. GENERAL

1. Aircraft conforms to FAA specifications
2. All AD notes complied with
3. Manufacturers' service letters, bulletins, instructions complied with
4. Aircraft papers in proper order
   a. Registration
   b. Airworthiness Certificate
   c. Radio license
   d. Flight Manual

Signature of Mechanic or Inspector _____

        Certificate No. _____

* Items not required by FAA as part of 100-hour inspection are marked with a star.

Figure 3-1. Annual Inspection Form (6 of 6).

The engine should respond smoothly to power changes, and when the pitch is changed on the propeller, it should be smooth and the recovery should be within the time limit allowed.

The engine should idle smoothly, and when the mixture control is pulled into the cut-off position, there should be a slight RPM rise before the engine dies.

The alternator or generator should show a current output, and all of the engine instruments should indicate that the systems are properly functioning. If the engine is fuel injected, the injector pressure should be within the range specified in the latest manufacturer's manuals. After the engine is thoroughly checked, the airplane is brought into the hangar and the cowling removed. The oil is drained while it is hot and one spark plug is removed from each cylinder to allow a differential compression test to be made.

(3) Compression Check

Remove one spark plug from each cylinder and perform a compression check. Either a direct or a differential check may be made, but the differential check is more uniform and is used in almost all shops. This type of check is not only quantitive, but if there is any trouble in the engine, it makes it easy to locate.

- Turn the propeller until the piston in number one cylinder is moving outward on its compression stroke. This stroke may be identified by air blowing out of the spark plug hole.

- Screw the adapter into the open spark plug hole and attach the hose of the compression tester to the adapter.

- Open the shut-off valve and turn the regulator in until there is between 10 and 15 PSI regulated air pressure in the cylinder.

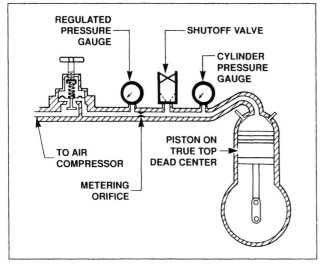

Figure 3-2. Differential compression tester

- Slowly move the propeller in the direction of normal rotation to bring the piston to top center. This position is indicated when the air pressure in the cylinder no longer pushes back on the piston. If the piston is moved too far, the air pressure will push it down, and the propeller should be brought back so the piston is well below top center, and then start back up. The reason for this is to have the rings properly seated in their grooves so they will provide their best seal.

Now that the piston is at top center, the air pressure can be increased to 80 PSI, and the cylinder pressure gage read.

CAUTION: With 80 PSI air in the cylinder, the propeller must not be bumped or the piston will be moved off of top center and the air will force it down, spinning the propeller fast enough to create a serious hazard.

The difference between the reading of the two gages indicated the amount of leakage in the cylinder. If the cylinder gage reads less than 60 PSI, it indicates a leakage of more than 25% and the cause of the leakage must be found and corrected. Listen for the escaping air to locate the problem. If air is heard coming from the crank case breather or the oil filler, the problem is air leaking around the piston rings. If the noise is heard at the exhaust pipe, the problem is exhaust valve leakage, and if the intake valve is leaking, the sound will be heard at the carburetor air inlet.

After the test has been made, reduce the air pressure with the regulator, and remove the adapter. Testing the next cylinder in firing order will speed up the test, as it will eliminate having to locate the compression stroke each time.

(4) Lubrication System

- While the oil is hot from the pre-inspection run up, remove the drain plug and drain the oil.

- Remove the oil filter and cut the element apart so you can examine the contaminants it has trapped. Any metal particles in the oil is an indication of an impending engine failure.

- Install a new filter element, and properly torque the filter in place.

- Reinstall and safety the drain plug and put the proper amount of the correct oil into the engine.

- Check all of the oil lines, the oil cooler, and the entire engine for any indication of oil leakage that might indicate a cracked crankcase or sump, or a leaking oil seal or gasket.

(5) Ignition System

One spark plug has already been removed from each cylinder for the compression check, and each of these should have been placed in a rack having numbered holes to indicate the cylinder from which they came. Remove the other plugs from the cylinders and examine all of them. Much of the condition within the cylinders can be determined by studying the spark plugs. Normal operation is indicated by a spark plug having a relatively small amount of light brown or tan deposits on the nose of the center electrode insulator. If there is an excess of deposits in the firing end cavity, and if these deposits are gray and clinker-like, lead fouling is indicated which may be caused by using a fuel with a higher than recommended tetraethyl lead content. Replacing the spark plug with one having a hotter heat range may reduce the fouling, but be sure that the hotter plug is specifically approved for the engine.

A dry, black, soot-like deposit in the spark plug indicates that the engine has been operating with an excessively rich mixture, and the induction system should be carefully checked for obstructions in the filter or for a malfunctioning carburetor heat valve.

If the back deposits are oily, there is a good probability that the valve guides are worn excessively, or there may be a broken piston ring. A brown shiny glaze on the nose insulator of the spark plug could indicate silicon contamination, and calls for a careful inspection of the carburetor air filter for air leaks around the filter element or for holes in the element itself. Any unfiltered air leaking into the induction system will allow sand or dust to enter the engine, and the intense heat in the combustion chamber will turn the silicon into a glass-like contaminant that, while being an insulator at low temperature, will become conductive at high temperatures and cause the spark plug to fail to fire when it is most needed.

Any spark plug whose electrodes have worn away to one-half of their original dimensions is considered to be worn out and should be replaced. If the spark plugs are not worn out, they may be reconditioned and reinstalled. Remove all of the lead deposits with a vibrator-type cleaner and, then, very lightly blast the firing end cavity with an approved abrasive. The ground electrode is then carefully moved over with a spark plug gapping tool to get the proper gap distance between the ground and the center electrode. Clean the inside of the terminal cavity with trichlorethylene and test the spark plug for operation. If the plug fires consistently under pressure in the tester, it may be returned to the engine. Use a new gasket and only a small amount of the thread lubricant recommended by the engine manufacturer. All of the spark plugs should be returned to the cylinder next in the firing

order to the one from which they were removed and should be swapped bottom to top.

The threads in the engine should be clean enough that the spark plug can be screwed down against the gasket with the fingers only, and then the spark plug is tightened to the torque recommended by the engine manufacturer, using a torque wrench of known accuracy.

Wipe the spark plug lead terminal free of all fingerprints with a rag dampened with trichlorethylene, and insert the lead straight into the spark plug and tighten the lead nut to the torque recommended by the manufacturer.

Check the magneto-to-engine timing using a timing light, and inspect the condition of the breaker points and the inside of the breaker compartment for any indication of moisture or oil. Be sure the ignition switch connection, the Plead, is tight and the wire is secure and not worn or chafed.

Be sure there are no damaged spark plug leads, and that all of the leads are properly supported.

(6) Fuel System

- Clean the main fuel screen, and the screen in the carburetor or fuel injection system.

- Replace them, using a new gasket, and after testing them under pressure for any indication of leaks, they must be safetied.

- Check all of the controls, and lubricate them as specified by the airframe manufacturer, being sure to use the proper lubricant.

- If the engine is fuel injected, check the manufacturer's recommendations for cleaning the injector nozzles, and if they are to be cleaned, be sure to follow the recommended procedure in detail.

- Check the entire fuel system for any indication of dye stains that would indicate a fuel leak.

(7) Induction System

- Remove and clean, or replace the induction air filter and check the entire system for any indication of leakage or deformation.

- Check the carburetor heat or alternate air valve to be sure that it does not allow any unfiltered air to enter the engine.

- Be sure that the valve and alternate air doors open fully so the induction airflow into the engine will not be restricted.

- The flexible hose that attaches the induction air valve to the hot and cold air sources should be carefully checked to be sure there are no kinks, and that the hose is in good condition with no possibility of its collapsing.

(8) Exhaust System

- Remove the shroud from around the muffler and check it for any indication of leakage. If there is an AD or manufacturer's service bulletin requiring pressure testing, be sure that this is complied with in detail — even a pinhole leak will fill the cabin with deadly carbon monoxide gas.

- Be sure that the cabin heat valve operates freely and has no obstructions in either the valve or the hose carrying heated air into the cabin.

- Check all of the cylinder heads at the exhaust port for indications of blown gaskets and for any indication of cracks or other leakage.

- Be very careful to check the exhaust pipes where the exhaust gas temperature probes enter them, as this is a possible source of leakage.

(9) Turbocharger

If the engine is equipped with a turbocharger, be sure to follow the inspection details specified by the manufacturer, as each installation has peculiarities that make them different.

And extreme care should be exercised to follow the manufacturer's recommendations in detail.

- The exhaust portion, including the operation of the waste gate, should be checked, as well as the induction air section which includes any relief valves, intercoolers, or manifold pressure sensors.

- The lubrication system and the mounting should be checked carefully, as some of these small turbines spin at speeds in excess of 100,000 RPM while operating red-hot.

(10) Cooling System

- Check all of the cooling system for cracks or damage and all of the baffles and seals that direct air through the fins on the cylinders.

- The cowl flaps should be checked for security and for full travel, both in their open and closed positions.

(11) Electrical System

Here again, systems vary between models of airplanes, and the manufacturer's recommendations must be followed in detail. Follow the manufacturer's recommendations regarding the inspection of brushes, bearings, and commutators of the starter and generator.

- The alternator and its mounting should be checked for security and for any indication of vibration-induced cracks.

- The voltage regulator and any relays or solenoids in the system should be checked for security and for tightness of all wires attached to them.

(12) Accessories and Controls

- Check all of the air, fuel, and hydraulic pumps for indication of leakage and for the condition of their seals.

- Check the vent and breather lines on all of the pumps, and the crankcase breather lines for security and for the proper positioning of their lower ends.

- Check the vacuum system to be sure that all filters in the pump inlet or the regulator are changed according to the manufacturer's recommendations, and the oil separator shows no indication of malfunctioning.

- Check the condition of the firewall to be sure that all controls, lines, and wires passing through are properly sealed, and there is no corrosion or other indication of damage.

- Check all controls for freedom and proper travel. The stops on the engine component should be reached just before the stop in the cockpit.

- Be sure that all of the shock mounts are in good condition, and the mounting bolts are properly torqued. It is important that the electrical grounding strap connecting the engine to the airframe be in good condition since the engine is mounted in rubber and all of the return current from the starter must flow through this strap.

(13) Propeller

The extreme stresses that the propeller encounters make it important that it be inspected very critically.

- Any nicks, cracks, or scratches in the blade must be carefully stoned out, or burnished, and any questionable area must be checked with one of the approved nondestructive test methods. Be sure that the blades are secure in the hubs, and that there is no oil leakage.

- Check the spinner and spinner bulkhead for any indication of cracks or damage, and check the governor for security and for full travel of its control.

(14) Post-inspection Run-up and Records

After the inspection has been completed, the engine is washed down, recowled, and checked out with a post-inspection run-up. When it checks out satisfactorily, the engine maintenance records must be completed, and all of the shop records filled out. The purpose of the 100-hour or annual inspection is to determine that the engine is in the condition required for its certification. There must be no alterations or

Ref. No.	Chart No.	Description	Serviceable Limit	New Parts Min.	New Parts Max.
30	1	First piston ring in cylinder (P/N 635814) . . . . Gap:	0.059	0.033	0.049
30	1	First piston ring in cylinder (P/N 639273) . . . . Gap:	0.074	0.048	0.064
31	1	Second piston ring in cylinder (P/N 635814) . . . . Gap:	0.050	0.024	0.040
31	1	Second piston ring in cylinder (P/N 639273) . . . . Gap:	0.069	0.043	0.059
32	1	Third piston ring in cylinder . . . . Gap	0.059	0.033	0.049
33	1	Fourth piston ring in cylinder . . . . Gap:	0.050	0.024	0.040
34	1	Fifth piston ring in cylinder . . . . Gap:	0.059	0.033	0.049
35	1	Piston pin in piston (standard or 0.005″ oversize) . Diameter:	0.0013L	0.0001L	0.0007L
36	1	Piston pin in cylinder . . . . End Clearance:	0.090	0.031	0.048
37	1	Piston pin in connecting rod bushing . . . . Diameter:	0.0040L	0.0022L	0.0026L
38	1	Bushing in connecting rod . . . . Diameter:		0.0025T	0.0050T
39	1	Connecting rod bearing on crankpin . . . . Diameter:	0.006 L	0.0009L	0.0034L
40	1	Connecting rod on crankpin . . . . Side Clearance:	0.016	0.006	0.010
41	1	Bolt in connecting rod . . . . Diameter:		0.0000	0.0018L
42	1	Connecting bearing and bushing twist or convergence per inch of length . . . . :	0.001	0.0000	0.0005
		CRANKSHAFT			
43	2	Crankshaft in main bearings . . . . Diameter:	0.005 L	0.0012L	0.0032L
44	2	Propeller reduction gear shaft in bearing . . . . . . Diameter:		0.0012L	0.0032L
45	2	Propeller drive shaft in shaft . . . . Diameter:		0.0012L	0.0032L
46	2	@Crankpins . . . . Out-of-Round:	0.0015	0.0000	0.0005
47	2	@Main journals . . . . Out-of-Round:	0.0015	0.0000	0.0005
48	2	Propeller drive shaft . . . . Out-of-Round:	0.002	0.0000	0.002
49	2	Propeller drive shaft in thrust bearing . . . . . End Clearance:	0.020	0.006	0.0152
50	2	Crankshaft run-out at center main journals (shaft supported at thrust rear journals) full indicator reading . . . . :	0.015	0.000	0.015
51	2	Propeller shaft run-out at propeller flange (when supported at front and rear journals) full indicator reading . . . . :	0.003	0.000	0.002
52	2	††Damper pin bushings in crankcheek extension . . . Diameter:		0.0015T	0.003 T
53	2	††Damper pin bushing in counterweight . . . . Diameter:		0.0015T	0.003 T
54	2	Damper pin in counterweight . . . . End Clearance:	0.040	0.001	0.023
55	2	Alternator drive gear on reduction gear . . . . . . Diameter:		0.001 T	0.004 T
56	2	Crankshaft gear on crankshaft . . . . Diameter:		0.000	0.002 T
		CAMSHAFT			
57	2	Camshaft journals in crankcase . . . . Diameter:	0.005 L	0.001 L	0.003 L
58	2	Camshaft in crankcase . . . . End Clearance:	0.014	0.005	0.009
59	2	Camshaft run-out at center journals (shaft supported at end journals) full indicator reading . . . . :	0.003	0.000	0.001
60	2	Camshaft gear on camshaft flange . . . . Diameter:		0.0005T	0.0015L
61	2	Governor drive gear on camshaft . . . . Diameter:	0.006 L	0.0005L	0.002 L
		CRANKCASE AND ATTACHED PARTS			
62	2	Thru bolts in crankcase . . . . Diameter:		0.0005T	0.0013L
63	1	Hydraulic lifter in crankcase . . . . Diameter:	0.0035L	0.001 L	0.0025L
64	2	Governor drive shaft in crankcase . . . . Diameter:		0.0014L	0.0034L

@   *If crankshafts are worn beyond these limits, they may be repaired by grinding crankpins and journals to 0.010 ″ under new shaft limits and renitriding the crankshafts. Crankshafts may be returned to factory for such repair.*

††   *Refer to Section 7-17 for allowable wear at damper pin bushings.*

Figure 3-3. Excerpts from table of limits for an aircraft engine.

modifications that have not been approved, and the approval data included with the engine records.

2. Overhaul

a. Top Overhaul

Most of the wear in a reciprocating engine occurs in the cylinder and piston assemblies, and if excessive oil consumption occurs, or if there is a loss of compression, the cylinders may be removed, the valves ground, and the piston rings replaced. This will usually restore the performance to the engine, and this procedure is called a top overhaul; since only the top of the engine is affected, the crankcase is not being opened.

b. Major Overhaul

One of the reasons an engine needs to be overhauled is to replace parts that have been worn beyond acceptable limits, and the engine manufacturer in his overhaul manuals gives two sets of limits. In figure 3-3, we have a typical table of limits. When the engine was originally built, it was assembled to the "New Parts" limits. For example, the crankshaft of a new engine should have a clearance between the shaft diameter and the bearing of between twelve ten-thousands of an inch (0.0012"), and thirty-two ten-thousandths of an inch (0.032"). If the engine is overhauled to new parts tolerances, this clearance must be maintained. But there is another set of limits given, "serviceable limits", and for the same crankshaft clearance, we see that this is five-thousandths of an inch (0.005"). The crankshaft can wear until its clearance in the bearings is five-thousandths of an inch and still be considered serviceable. If the overhaul is done to serviceable limits, the time between overhaul will be considerably shortened, and the part will definitely have to be replaced on the next overhaul.

All major overhauls must be done in strict accordance with the manufacturer's overhaul manual, and all service bulletins and Airworthiness Directives must be complied with. A typical major overhaul will be discussed in this book to familiarize you with what is done when an engine is major overhauled.

3. Remanufacture

There is little technical difference between an engine that is overhauled and one that is remanufactured, but the FAA does make provision for the engine manufacturer or an agency approved by the manufacturer to issue zero time records to an engine that has been remanufactured by them. To quote FAR 91.175, "For purposes of this section, a rebuilt engine is a used engine that has been completely disassembled, inspected, repaired as necessary, reassembled, tested, and approved in the same manner and to the same tolerance and limits as a new engine with either new or used parts. However, all parts used in it must conform to the production drawing tolerances and limits for new parts, or be of approved oversized or undersized dimensions for a new engine." If an overhaul facility other than the manufacturer or a shop approved by him does exactly the same type of overhaul, they cannot issue a zero time record. New records can be started with *zero time since major*, but the total time on the engine must be included. When an engine manufacturer rebuilds or remanufactures an engine, the engine actually loses its identity. Those that are returned to the factory are disassembled, cleaned, inspected, and all components repaired as necessary. The parts are then put into bins, and engines are assembled from this stock of reconditioned parts. Fits and tolerances used are the same as those for new engines.

Study Questions

1. *Must an aircraft engine be overhauled at the manufacturer's recommended TBO?*

2. *How does a manufacturer determine his recommended TBO?*

3. *What is meant by a top overhaul?*

4. *When is it economical to top overhaul an aircraft engine?*

5. *What should be done if an engine is operated beyond its recommended TBO?*

6. *What damage can be done to an engine by operating with a fuel having a tetraethyl lead content higher than that for which the engine is designed?*

7. *What attritional force can do more damage to an engine when it is not run than when it is operating every day?*

8. *Why is it considered bad practice to operate an engine for only a ground run-up and then allow it to sit for extended periods of time?*

9. What should be done to the inside of the cylinders of an engine that is not to be operated for an extended period of time?

10. List six items that should be checked on a good preflight inspection of an aircraft engine.

11. Is a 50-hour inspection required for an aircraft engine?

12. Who is authorized to conduct an annual inspection on an aircraft engine?

13. What are the two types of compression tests that can be used on an aircraft engine?

14. What would likely be indicated by a reading of 55/80 on a differential compression test if there was the sound of air escaping at the exhaust stack?

15. What should be done to the oil filter element that has been removed on an annual inspection?

16. What would likely be indicated by a black, oily deposit on the spark plugs removed from one of the cylinders?

17. Would you install a spark plug having a hotter or a colder heat range if all of the spark plugs removed from an engine had excessive lead deposits in their firing end cavity?

18. What may be properly used to clean the spark plug lead terminal before putting it into the spark plug?

19. Why is it extremely important that there be absolutely no leaks in an exhaust system muffler?

20. After spark plugs have been serviced, should they be returned to the same hole from which they were removed?

21. What are two sets of limits to which an engine can be overhauled?

22. Who is authorized to remanufacture an aircraft engine and return it to service with zero time records?

Chapter IV

Major Overhaul

A. Removal From The Airplane

We will follow a typical engine overhaul through a shop to better understand what is done. When the airplane is brought into the shop for engine overhaul, the proper work order is started and a record is made of the serial number of all of the components that are to be kept with the engine. When the engine is removed and sent to the engine overhaul shop, it normally includes the intercylinder baffles, the carburetor or fuel injection system, the magnetos, ignition leads and spark plugs, and the induction system from the carburetor to the cylinders. The exhaust system, vacuum pump, hydraulic pump, propeller and its governor, and most other accessories are not considered to be part of the engine and are sent to specialty shops for their overhaul; or, if this work is done by the same shop as is overhauling the engine, they are run through with separate paperwork. The oil is drained from the engine and the propeller is removed, then the engine is stripped of all but the accessories going with it, and it is then taken to the engine overhaul shop.

B. Disassembly

The engine is mounted on a stand and the magnetos are removed and sent to the proper department for their overhaul, and the carburetor or fuel injection system is removed and either sent out, or carried to the proper department for its overhaul. The engine is completely disassembled and laid out in an orderly fashion so it can be given its preliminary inspection.

The condition of all of the parts are noted and a record made if any are obviously damaged, so this damage will not be overlooked as the part progresses through the overhaul process.

C. Cleaning

After all of the parts have been disassembled, they are degreased by soaking or spraying them with some form of mineral spirit solvent such as varsol or white furnace oil. Water-mixed degreasing compounds usually contain some form of alkali which, if allowed to remain in the pores of the metal, will, when the engine is returned to service, react with the hot lubricating oil to form soap and cause foaming, resulting in failure of the lubricating system.

Degreasing will remove dirt, grease, and soft carbon, but many parts will have deposits of hard carbon on their interior surfaces and all of this must be removed. An approved decarbonizing solution must be used following the manufacturer's recommendation in detail. These solutions are generally quite active and are often heated, so care must be taken that no parts are left in the solution longer than absolutely necessary to loosen the carbon deposits, and that no magnesium parts are placed in the solution unless it is known *not* to react with magnesium.

After the parts have been removed from the decarbonizing vat, they are thoroughly cleaned of all traces of the solution with a blast of wet steam or by brushing them with mineral spirits.

Any hard carbon that was not removed by the decarbonizing solution may be removed by dry blasting with plastic pellets or with such organic materials as rice, baked wheat, or crushed walnut shells. Be sure that all machined surfaces are masked off and all passages are plugged or covered. Use as low a pressure as practical and blast only enough to remove the carbon.

Parts may be vapor-grit blasted. However, care must be exercised that only small grit be used and adequate protection taken to prevent any of it remaining in the part after it is cleaned.

When cleaning pistons, care must be used to prevent scratching their highly stressed surfaces. Do not scrape off the carbon deposits, but remove them by solvent action or by soft grit or vapor blasting, and especially, do not use automotive-type ring groove cleaners as it is critical that the bottom radius of the groove be maintained. Soften the carbon and remove it by drawing binder twine through the groove. Many aircraft pistons are cam ground; that is, they are not perfectly round but are slightly oblong so they will round out as they expand from the heat. It is important that the skirt of the cam-ground piston is not abraded in the cleaning process.

All bearing surfaces must be polished with crocus cloth moistened in mineral spirits, and afterward with dry crocus cloth. All passages in the crankcase must be thoroughly cleaned by flushing them with wet steam, and following this with a spray of mineral spirits.

After all of the parts are thoroughly cleaned and the steel parts coated with a film of protective oil to prevent their rusting, they are ready to be inspected.

D. Inspection

Before starting the inspection process, an inspection and overhaul form should have been started, and must be with the engine at this point, along with the manufacturer's overhaul manual and all of the proper inspection equipment. Throughout the overhaul, the engine manufacturer's manuals, bulletins, and other service information must be available. They are the final authority on the suitability of a part for reuse.

1. Visual Inspection

Inspect all parts under a good light for any indication of surface damage such as nicks, dents, deep scratches, visible cracks, distortion, burned areas, pitting, or pick-up of foreign metal. Inspect all studs for indication of possible bending, looseness, or partial removal. All threads should be free from nicks or other forms of damage.

The visual inspection should allow all of the parts to be separated into three groups: those which should be discarded, those which require rework, and those which are apparently serviceable.

2. Fluorescent Penetrant Inspection

All of the nonferrous parts that have been determined to be apparently serviceable are further inspected by a fluorescent penetrant inspection method. This procedure checks for surface cracks and discontinuities that could escape detection by the naked eye.

The most important requirement for a good fluorescent penetrant inspection is that the part being inspected be absolutely clean with all contamination removed not only from the surface but especially from any cracks that may exist. For best results, the part should be vapor-degreased with hot trichlorethylene vapors. While it is still hot, it should be immersed in the fluorescent penetrant solution and allowed to stay long enough for the solution to thoroughly soak into all cracks. After the required dwell time, as this soaking is called, the part is removed and all of the penetrant on the surface flushed off with a soft spray of hot water, and the part baked dry in an oven. Now the clean, dry part is put into the developer tank where a fluffy talcum-like powder covers it and pulls the penetrant out of any cracks that may exist in the metal. The part is then inspected under a black light where any crack extending to the surface will appear as a green line.

The interpretation of the results of the fluorescent particle inspection is its most critical aspect. Knowing exactly where to look for cracks and being able to recognize a true crack from a false indication can best be learned from an experienced NDI technician.

3. Magnetic Particle Inspection

Ferrous parts (those containing iron) can be inspected for hidden cracks by magnetic particle inspection in which the part is magnetized, and then a fluid containing particles of iron oxide treated with a fluorescent dye is flowed over the part. A crack will interrupt the flow of magnetic flux within the part

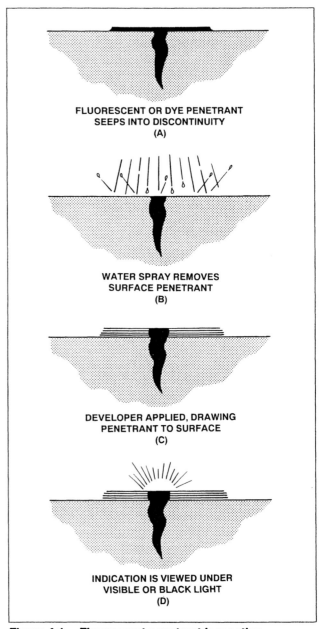

FLUORESCENT OR DYE PENETRANT
SEEPS INTO DISCONTINUITY
(A)

WATER SPRAY REMOVES
SURFACE PENETRANT
(B)

DEVELOPER APPLIED, DRAWING
PENETRANT TO SURFACE
(C)

INDICATION IS VIEWED UNDER
VISIBLE OR BLACK LIGHT
(D)

Figure 4-1.   Fluorescent penetrant inspection

and produce magnetic poles on the surface of the metal: These poles will attract the iron oxide. When the part is inspected under a black light, any crack present will appear as a brilliant green line.

There are two ways a part can be magnetized. The engine overhaul manual specifies the method to be used. Figure 4-2 is a page from an engine overhaul manual specifying the methods. For example, the crankshaft must be magnetized both circularly and longitudinally to get an indication of any cracks that might appear. Figure 4-3 shows the methods of getting both types of magnetization. Figure 4-4 shows the way each type of magnetization will indicate a crack.

When checking for cracks that run across the part, the part is magnetized longitudinally; that is, it is placed in a coil, or solenoid, and current flowing through the coil will magnetize the part with the poles at its ends. Now, any crack that runs across the part will show up by attracting the oxide. After inspecting the part for transverse cracks, as these are called, the part is completely demagnetized and remagnetized with circular magnetization. For this magnetization, the part is clamped tightly between the heads of the machine and the current is passed through the part itself. The lines of magnetic flux radiate out from the part and will magnetize it with no external poles. If a crack is present, it will interrupt the flux and poles will appear, attracting the oxide. Small cylindrical parts such as nuts, washers, springs, and pins may be circularly magnetized by placing them on a conductive rod clamped between the heads of the machine. It is extremely important that after a part is magnetically inspected, every trace of the magnetism be removed. This may be done by placing the part in a coil through which alternating current is passed, rotating the part back and forth, and completely removing it from the field while the current is flowing. A better method of demagnetizing is to use pulses of direct current whose polarity is reversed with each pulse and the intensity of the pulses gradually decreased to zero. Either of these two methods disorient the magnetic field and demagnetize the part.

4. Dimensional Inspection

Use the proper measuring tool such as micrometer calipers, telescoping gages, and dial indicators to measure the parts and determine that the fits are as

Part	*Method of Magnetization	D.C. Amperes	Critical Areas	Possible Defects
Crankshaft	Circular and Longitudinal	2500	Journals, fillets, oil holes, thrust flanges, prop flange.	Fatigue cracks, heat cracks.
Connecting Rod	Circular and Longitudinal	1800	All areas.	Fatigue cracks.
Camshaft	Circular and Longitudinal	1500	Lobes, journals.	Heat cracks.
Piston Pin	Circular and Longitudinal	1000	Shear planes, ends, center.	Fatigue cracks.
Rocker Arms	Circular and Longitudinal	800	Pad, socket under side arms and boss.	Fatigue cracks.
Gears to 6″ Diameter	Circular or on Center Conductor	1000-1500	Teeth, Splines, Keyways.	Fatigue cracks.
Gears over 6″ Diameter	Shaft Circular Teeth Between Heads Two Times 90°	1000-1500	Teeth, Splines.	Fatigue cracks.
Shafts	Circular and Longitudinal	1000-1500	Splines, Keyways, Change of Section.	Fatigue cracks, heat cracks.
Thru Bolts, Rod Bolts	Circular and Longitudinal	500	Threads Under Head.	Fatigue cracks.

Note: (*)

LONGITUDINAL MAGNETISM: Current applied to solenoid coil surrounding the work.

CIRCULAR MAGNETISM: Current passed through work or through non-magnetic conductor bar inserted through work.

Figure 4-2. Typical specifications for current requirements for magnetic particle inspection.

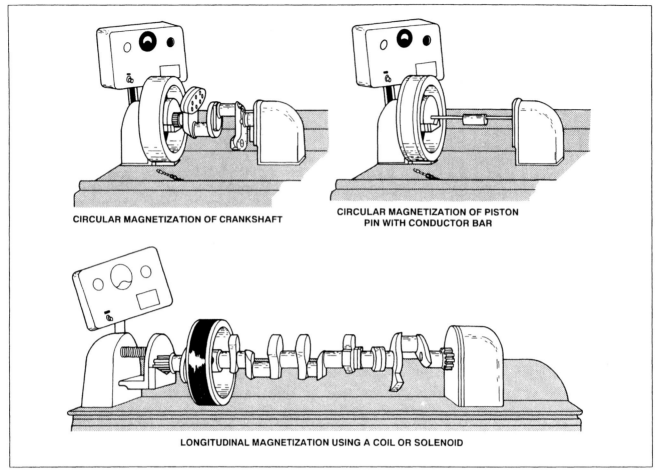

CIRCULAR MAGNETIZATION OF CRANKSHAFT

CIRCULAR MAGNETIZATION OF PISTON PIN WITH CONDUCTOR BAR

LONGITUDINAL MAGNETIZATION USING A COIL OR SOLENOID

Figure 4-3. Magnetic particle inspection.

specified in the engine overhaul manual. You will notice in figure 3-3 that both new parts limits and serviceable limits are given. It is poor economy to overhaul an engine to serviceable limits, as it will run only a short time before it will be outside of these limits. The increased clearances will accelerate wear and decrease the time between overhauls. Also, when the engine is next overhauled, it will most probably be more expensive than it should be because of the increased part replacement caused by these excessive clearances.

In order to measure the crankshaft clearances we are using in our example, new main bearing inserts are installed in the crankcase halves, and the case is assembled and torqued to the values recommended in the table of torques. Telescoping gages are adjusted to the inside diameter of the bearings and are then measured with micrometer calipers. The journals of the crankshaft are carefully measured with the same micrometer caliper. The difference between the two measurements is the clearance between the journal and the bearing. For a new engine, this fit is allowed to between 0.0012L and 0.0032L. The "L"

following this dimension indicates that the fit is loose, meaning that the bearing inside diameter is larger than the outside diameter of the journal.

The service limit of this fit is 0.005L. A crankshaft whose fit falls outside of this limit, or one with journals more than a half thousandth inch (0.0005") out of round can be ground as much as 0.010" undersize, its surface renitrided and then fitted into the case with undersize bearing inserts.

All of the dimensions called for in the table of limits must be measured. If the part does not fall within the desired tolerance, the parts must be replaced.

Some fits, such as the bushing in the small end of the connecting rod, are called interference fits and in our example are dimensioned as 0.0025T to 0.0050T. This means that the bushing must be from two and one-half to five thousandths of an inch larger than the hole it fits. To press this bushing into place, an arbor press and a special bushing installation drift is used to force the bushing into the rod. Other interference fits, such as the valve guides and valve seats in the cylinder heads, are so tight that the head must be heated in an oven and the guides

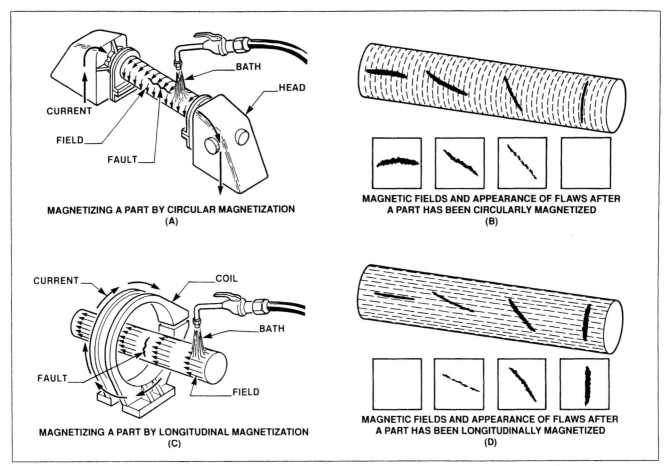

MAGNETIZING A PART BY CIRCULAR MAGNETIZATION
(A)

MAGNETIC FIELDS AND APPEARANCE OF FLAWS AFTER
A PART HAS BEEN CIRCULARLY MAGNETIZED
(B)

MAGNETIZING A PART BY LONGITUDINAL MAGNETIZATION
(C)

MAGNETIC FIELDS AND APPEARANCE OF FLAWS AFTER
A PART HAS BEEN LONGITUDINALLY MAGNETIZED
(D)

Figure 4-4. Magnetic particle inspection showing how each type of magnetization indicates a crack.

and seats chilled with dry ice and then assembled while they have large dimensional differences because of the temperature extremes.

E. Repair

If any of the inspection shows that parts need repairing, they must be brought back to the standards required by the engine manufacturer before the engine is reassembled.

1. Crankcase

Crankcases are subjected to such high stresses that cracks are likely to appear, and these must be repaired. Crankcases are quite expensive, and modern welding technology has made the welding of cracks an acceptable repair. It must be noted, however, that repairs of this nature must be done either by the engine manufacturer or by some facility with an VA repair station approval for this specialized type of work.

Welding is done by one of the forms of inert gas arc welding. After some of the weld material has been deposited, the bead is peened to relieve stresses built up by the welding. After the complete weld is made, the repair is machined to match the rest of the surface. Not only are cracks repaired by welding, but bearings have been known to turn in their saddles and damage the bearing cavity. These cavities may be repaired by building them up with weld material and line boring the case after all of the repairs have been made.

Camshafts normally run in bearings cut in the crankcase without any bearing inserts. If the clearance between the case and the camshaft is greater than is allowed, the bearings may be line-bored oversize and an oversize camshaft installed.

All of the studs must be checked for any indication of their having been loosened in operation. Any that are bent or loose must be replaced using oversize studs if necessary. Figure 4-5 shows the markings used to identify the standard oversize studs.

If the threads in the case are stripped out, they may be drilled clean with a special drill, the hole tapped with a special tap, and a stainless steel heli-coil insert screwed into place. Heli-coils provide new threads for the hole, and standard studs may be installed. There is no decrease in strength when this type of repair is made.

TYPICAL PART NO.	OVERSIZE ON PITCH DIA. OF COARSE THREAD (INCHES)	OPTIONAL IDENTIFICATION MARKS ON COARSE THREAD END		IDENTIFICATION COLOR CODE
		STAMPED	MACHINED	
XXXXXX	STANDARD	NONE		NONE
XXXXXXP003	.003			RED
XXXXXXP006	.006			BLUE
XXXXXXP009	.009			GREEN
XXXXXXP007	.007			BLUE
XXXXXXP012	.012			GREEN

Figure 4-5. Standard and oversize stud identification.

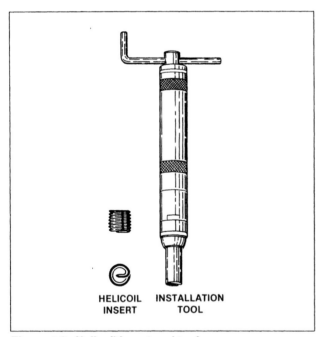

Figure 4-6. Helicoil insert and tool.

HELICOIL INSERT   INSTALLATION TOOL

2. Crankshaft

The crankshaft is the heaviest and most highly stressed part of an aircraft engine, and there are very few repairs that can be made to it. Those repairs which can be made must be done by either the manufacturer or by repair stations specially approved for this type of work.

If the journals are scored or worn out-of-round, the shaft can be ground to the proper undersize. When it has been determined by magnetic particle inspection that there are no cracks in the crankshaft and that the shaft is not bent, it is placed in a special lathe and all of the main and connecting rod journals ground to the proper size. Special care must be taken with the radius between the bearing surface and the crank cheek because of the extremely high stresses encountered with aircraft crankshafts. After all of the journals have been ground, they are polished and the crankshaft is surface-hardened by the nitriding process and the propeller flange area cadmium-plated. The crankshaft is given a final magnetic particle inspection and all of the sludge plugs, counterweights, and all other removable components are reinstalled.

3. Cylinders

Bent cylinder barrel fins may be straightened, and any cracked head fins may be dressed smooth. Be sure that no more fin area is removed than that allowed by the manufacturer as these fins are extremely important for proper cooling.

If the cylinder bore is worn beyond the allowable limits, it may be restored to its original dimensions

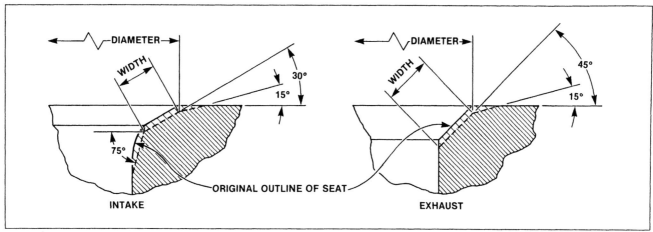

Figure 4-7. Cutting and narrowing valve seats.

by porous chrome plating done by a repair station approved for this work.

The valve guides may be replaced by pressing the old guides out and heating the cylinder head in an oven while chilling the new guide in dry ice and pressing the guide in place. New valve seats may be installed in the same way.

When the new seats and guides have been installed, the guides are reamed to size and the seats are ground using the guide as a pilot hole. The recommendations of the engine manufacturer must be followed in detail in the grinding of the seats, but a typical procedure for grinding a 30° intake valve seat would be to use a 30° stone, then a 15° stone, to cut the top of the seat to the diameter specified in the overhaul manual. A 75° stone is then used to narrow the seat to the width specified.

4. Valves and Valve Springs

Some valves may not be reused, and since the valve is the part in the engine subject to the most severe wear, those that may be reused must be carefully examined for any indication of overheating that could render them unserviceable.

Any nicks or scratches in the valve stem near the valve spring retainer groove is cause for rejection of the valve. Any valve whose stem diameter in the center measures less than the diameter at the spring end should be rejected as this is an indication that the valve has stretched.

Chuck the valve in the valve grinding machine, and, using a dial indicator, check to see that the face runs true with the stem within a very small tolerance, usually somewhere around one and one-half thousandths of an inch (0.0015″). Remove only enough material to clean up any wear marks or pits on the valve face, and be sure that the surfaced valve has at least the minimum edge thickness specified in the overhaul manual.

If the overhaul manual specifies an interference fit be ground between the valve face and the valve seat, the valve face is ground between one-half and one degree flatter than the seat so the valve will seat with a line contact on its outer edge.

Figure 4-8. Checking valve for stretch and damage.

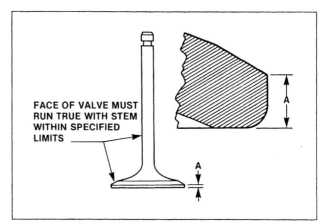

Figure 4-9. Valve face measurement.

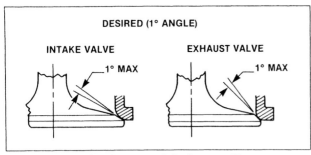

Figure 4-10. Valve face-seat interference fit.

Check the valve springs with a spring tester to be sure that they require the proper force to compress them to the specified height.

5. Pistons and Rings

After the pistons have been cleaned and all of the carbon removed from the ring grooves and the oil relief holes in the lower ring groove, the piston is dimensionally checked and inspected for any indication of overheating or scoring. A new set of the proper piston rings is installed after first measuring the end gap by placing the ring in the cylinder barrel and squaring it up with the piston and then using a feeler gage to measure the end gap. If the gap is correct, the rings may be installed on the piston with the part number toward the top of the piston. Use a ring expander and be careful that the piston is not scratched by the ends of the ring. With the rings installed, check the clearance between the ring and the side of the ring groove. If tapered rings are installed, a straight edge should be held so the ring is flush with the side of the piston and a measurement made with the feeler gage.

It is extremely important that only piston rings approved for the engine be used, and that the ring be compatible with the cylinder wall. Only cast-iron

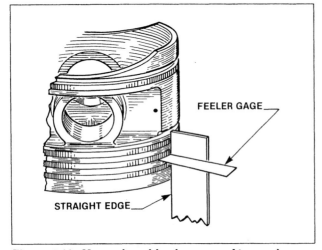

Figure 4-11. Measuring side clearance of tapered piston rings.

rings may be used with nitrided or chrome-plated cylinder walls, but chromed rings may be used in plain steel cylinders.

6. Connecting Rods

After the connecting rods have passed the magnetic particle inspection and a new small-end bushing has been pressed into place and reamed to size, a new bearing insert is installed in the large end and new rod bolts and nuts are installed. Arbors are installed in both ends of the connection rod, and a measurement is made using a parallelism gage to check for a bent rod, as would be indicated if the two arbors were not exactly parallel. With the arbors still installed, the rod is laid across parallel blocks on a surface plate and checked to see whether a feeler gage can be passed between the arbor and the block, which would indicate that the rod is twisted. If any rod is twisted or bent beyond the limits specified, it must be replaced.

When a new rod is installed, it should match the rod on the opposite side within one-half ounce to minimize vibration.

7. Camshaft

Since the camshaft is responsible for opening the valves at the proper time, all of the lobes must be examined to see that they are not excessively worn. If any of the hydraulic valve tappet bodies were spalled or pitted, the lobe that operated that valve must be inspected for any surface irregularities or feathering of the edge. If any such condition is found, the camshaft should be rejected. If the overhaul manual specifies that the lobes are tapered, a dimensional check should be made to indicate whether the amount of taper is within the limits specified.

If all of the lobes are in good condition, check the shaft for bends by supporting its end bearings in V-blocks and checking the center bearing for runout. Then check the bearing diameters and compare them with the bearing in the case for the proper fit.

8. Valve Operating Mechanism

Almost all modern aircraft engines use hydraulic valve lifters to maintain a zero clearance in the valve operating mechanism.

The valve tappet bodies that ride on the cam lobes are examined visually and dimensionally, and if they pass this, they are inspected by magnetic particle inspection and thoroughly demagnetized. The hydraulic plunger assembly must never be magnetized as magnetization will prevent the steel ball-type check valve from seating. The plunger and cylinder are matched units, and parts from one should never be interchanged with parts from

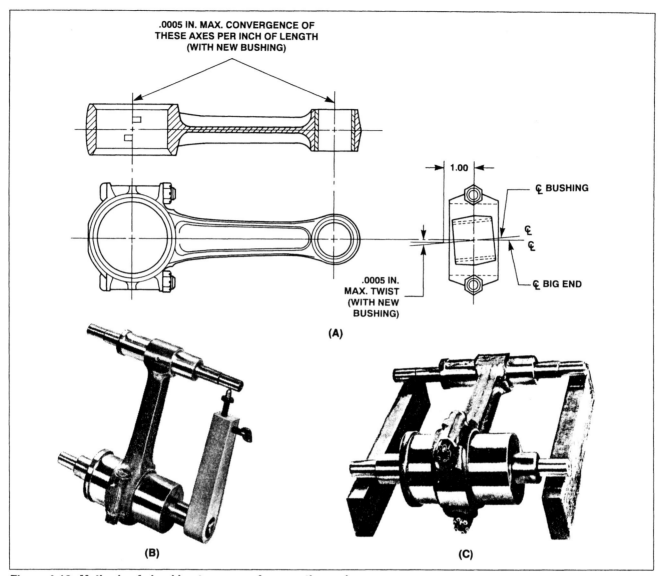

.0005 IN. MAX. CONVERGENCE OF
THESE AXES PER INCH OF LENGTH
(WITH NEW BUSHING)

1.00

₵ BUSHING

₵
₵

₵ BIG END

.0005 IN.
MAX. TWIST
(WITH NEW
BUSHING)

(A)

(B)

(C)

Figure 4-12. Methods of checking trueness of connecting rods.

another. After the part is thoroughly cleaned and visually checked for any chipped shoulders or evidence of other damage, it is given a leakage test by inserting the plunger into the cylinder and quickly depressing it. If it bounces back, it indicates that the check valve is seating properly and the assembly is satisfactory for reinstallation. If it does not bounce back, the valve is not seating and the unit must be replaced.

The push rods must be inspected for straightness by rolling them across a flat surface plate. The ball ends should be checked to be sure they are tight in the rod.

F. Reassembly

After all of the parts have been inspected and repaired as necessary, the engine is ready to be reassembled. The procedure here is typical, but *the actual assembly must be done in strict accordance with the manufacturer's overhaul manual.*

1. Cylinders

Lubricate the valve stems with the recommended lubricant, and insert the valves in the valve guides. Place the cylinder over a post and hold the valves in place while you slip the valve springs and retainers over each valve stem. Using the proper valve spring compressor, compress the valve springs and install the valve stem keys and any valve rotators that are used.

Slip a cylinder base seal around the skirt of the cylinder and install any of the inter-cylinder baffles or fin stabilizers that are required.

2. Pistons and Rings

Lubricate the piston, the rings, and the wrist pin with the appropriate lubricant, and stagger the ring gaps in the way specified in the overhaul manual. Slip the wrist pin into one side of the piston and, using the proper ring compressor, compress the piston rings and slip the piston into the cylinder up to the wrist pin.

3. Crankshaft

All of the sludge tubes and expansion plugs must be installed in the reconditioned crankshaft, all of the counterweights properly assembled, and any gears that are attached to the shaft properly secured and safetied. The crankshaft can now be mounted in a built-up fixture to hold it upright while the connecting rods are assembled to the shaft using new bolts and nuts. The rods are now torqued and safetied. Be very sure when assembling the rods to the crankshaft that the numbers stamped on the rods are on the side specified in the overhaul manual.

4. Crankcase

Place the crankcase halves on a flat work surface and install the main bearing inserts, being sure that the locking tangs or dowels are properly in place. Lubricate and install the hydraulic tappet bodies in one crankcase half, and then slip the camshaft in place after lubricating its bearing surfaces. Now the main bearings may be lubricated and the crankshaft may be installed. The front oil seal is installed around the propeller shaft, and a *very thin* coating of non-hardening gasket compound is applied to the

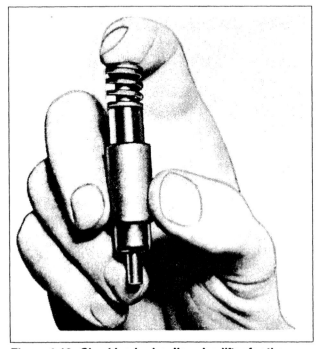

Figure 4-13. Checking hydraulic valve lifter for the condition of the check valve.

outside mating surface of each half of the crankcase. If the manufacturer recommends it, a very fine silk thread is embedded in the gasket compound on one of the crankcase halves.

Very carefully, holding the tappet bodies in place, lower the crankcase half over the one in which the camshaft and crankshaft are installed, being very careful that the front oil seal is properly seated. Any special instructions regarding the seating of the main bearings must be observed. Install new nuts on all of the studs and through-bolts and, using torque hold-down plates over the cylinder pads, torque all of the fasteners to the specifications in the overhaul manual, being very careful to use the *exact torque and sequence* specified.

The crankcase may again be installed in its vertical position in the built-up fixture and the gears in the accessory case installed with careful attention being paid to the timing of the cam gear to the crankshaft gear.

Now, the torque hold-down plates may be removed and the cylinders slipped in place, installing the proper push rods and push rod housings with the appropriate seals. The rocker arm and rocker shafts are lubricated and installed and the rocker shaft secured in the cylinder head.

5. Final Assembly

The oil sump is installed on the crankcase and the magnetos are installed and timed. The carburetor or fuel injection system and all of the induction system are installed, as well as all of the various accessories such as the pumps, the generator or alternator, and any baffles or other devices needed for the engine to operate properly.

G. Testing

1. Test Facilities

After an engine has been overhauled, it should be run-in and tested. Either a dynamometer such as the one in figure 4-14 or a special test club propeller should be installed to absorb the horsepower developed by the engine.

A sheet metal cooling air scoop should be installed on the engine to assure that all of the cylinders are properly cooled, and a blast of cooling air should be directed across the generator or alternator.

A clean induction air filter of sufficient size to prevent any obstruction to the airflow must be installed, and an exhaust system either of the type for which the engine installation is approved, or one specifically approved by the engine manufacturer must be installed.

The engine must be equipped with at least the following test instruments and, because of the inaccuracy of the instruments in most aircraft, the aircraft instruments should not be used.

1. An accurate tachometer
2. An accurate oil pressure gage
3. An accurate oil temperature gage
4. A cylinder head temperature gage that reads the temperature of the cylinder specified by the engine manufacturer
5. A water manometer to measure the pressure inside the crankcase
6. An accurate ammeter

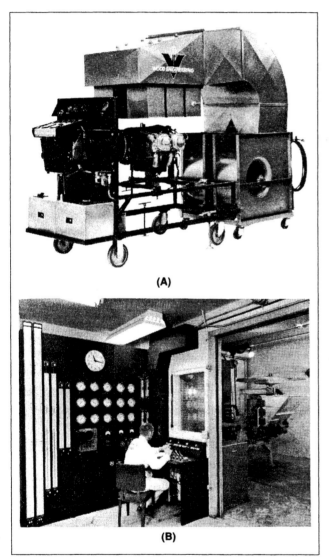

Figure 4-14.
- (A) A special dynamometer used for measuring the horsepower produced by an aircraft engine.
- (B) A freshly overhauled engine can be run-up on a special run-in stand and a test club propeller can be used to act as a load.

The test stand must be equipped with a fuel system that will deliver an adequate flow of clean fuel at the correct pressure.

2. Test Preparation

If the engine is equipped with a pressure carburetor, it must be allowed to sit with fuel in the fuel chambers for at least eight hours before the engine is run.

The engine must be pre-oiled. This may be done with a pressure oiler which forces oil through all of the oil passages in the engine until it runs out, indicating that all of the passages are full. Another way that is quite satisfactory for some of the smaller engines is to remove one of the spark plugs from each cylinder and spin the engine over with the starter until pressure is indicated on the oil pressure gage.

3. Test Run

After the engine is pre-oiled and it has been assured that all of the controls are operating properly, the engine may be started and run through a test schedule specified in the engine overhaul manual. A typical schedule is shown in figure 4-15. Part of this run is an oil consumption test. After the engine has operated for a specific amount of time under controlled conditions, all of the oil is drained and weighed, then is put back into the engine, and the oil consumption run is made, after which the oil is again drained and weighed.

4. Installation and Test Flight

When the ground run has been completed, the engine is installed in the airplane and all of the installation is carefully inspected. After the aircraft is approved for return to service by the appropriate mechanic, it must be test flown.

	MAJOR OVERHAUL TEST RUN GTSIO-520-C		
Time-Minutes	**RPM**		**T/C Outlet PR." Hg (Reference)**
5	1200		
10	1500		
10	2100		
10	2600		
10	2800		
10	3000		
5	3200± 25	100% Power	
5	*3000	82.3% Power - (280 BHP)	35.0 - 36.0
5	*2600	53.5% Power - (182 BHP)	34.7 - 35.7
10	600 ± 25	Idle Cooling Period	34.4 - 35.4

NOTE

Stop engine, drain oil, weigh oil for oil consumption determination and replace in engine.

Figure 4-15. Typical run-in schedule

It was at one time thought that aircraft engines should be *babied* and flown with very low power settings for quite some time to begin with, but this has been proven totally wrong. If an engine is treated in this manner, it is very likely that, rather than the rings seating in the cylinders and stopping the initial high oil consumption, a hard varnish will form on the cylinder walls and prevent the rings ever seating. Most engine manufacturers now recommend that the initial flights be made using standard power for climbs and 75% power for cruise. When making these initial flights, all of the climbs should be shallow so there will be sufficient airspeed to maintain enough air over the engine for proper cooling.

5. 25-Hour Inspection

After an engine has undergone a major overhaul, it is run-in with straight mineral oil because this allows the rings to seat to the cylinder walls better than the normally used ashless dispersant (AD) oil. After the engine has flown for twenty-five hours it should have this oil drained out, the filter replaced, and all of the strainers in both the fuel and oil system carefully checked for any indication of foreign matter.

The first time the oil is drained on a new or freshly overhauled engine, there will be some very fine metal shavings and lint in the filters; but this is normal, as the parts wear themselves together. It is chunks of metal that are of real concern.

Study Questions

1. *What accessories are normally returned with an engine when it is sent in for overhaul?*

2. *Why should special caution be exercised if aluminum parts are cleaned with water-mixed degreasing compounds?*

3. *What is used to remove carbon deposits from the inside of an aircraft engine crankcase?*

4. *What may be used to safely clean the ring grooves in an aircraft piston?*

5. *How should a part be cleaned if it is to be inspected by the fluorescent penetrant method?*

6. *What are two methods of magnetizing a part for magnetic particle inspection?*

7. *Why is it poor economy to overhaul an engine to serviceable limits?*

8. *How is the clearance between the crankshaft journal and its bearing measured?*

9. *What is meant by an interference fit?*

10. *Who is authorized to repair an aircraft crankcase by welding?*

11. *What may be done to a crankcase if the camshaft bearings have worn oversize?*

12. *What is indicated by a conical end machined onto a replacement stud?*

13. *What must be done to a crankshaft after it has been ground undersize?*

14. *What may be done to restore a cylinder to its original dimensions after it has been worn beyond its allowable limits?*

15. *How is a valve seat ground to the proper width?*

16. *How much material should be removed from a valve when it is being refaced?*

17. *If a manufacturer specifies that the valve be ground with an interference fit between the valve and its seat, which is ground to the flatter angle: the valve or the seat?*

18. *What kind of test is performed on a valve spring to determine whether or not it is suitable for reinstallation?*

19. *How is the end gap on a piston ring measured?*

20. *How is the side clearance measured on a wedge-shaped piston ring?*

21. *What type of piston ring would be proper for use in a chromed cylinder?*

22. *What kind of dimensional checks should be made of a connecting rod?*

23. *Why should the plunger assembly of a hydraulic valve lifter not be inspected by the magnetic particle method?*

24. *How is the hydraulic valve lifter checked for proper operation?*

25. *How are push rods checked for straightness?*

26. What may be used between the crankcase halves to prevent oil leaking from the case?

27. What consideration, besides the amount of torque, is necessary to consider when bolting the halves of a crankcase together?

28. What are two ways a freshly overhauled engine may be loaded for its test-stand run-in?

29. Why should the standard instruments installed in an aircraft not be used for the ground run-in?

30. Give two methods that may be used to pre-oil an aircraft engine.

31. How is oil consumption determined on a ground run-in?

32. What power should be used for the flight run-in of a freshly overhauled engine?

33. What kind of oil should be used for the initial break-in period of an aircraft engine?

Chapter V

Engine Troubleshooting

Efficient troubleshooting is based on a logical analysis of what is happening to be able to determine the cause of a malfunction. There is no magic to successful troubleshooting, but it is rather the application of cold logic and a thorough knowledge of the basics of engine operation.

When an experienced technician is called onto a problem dealing with deteriorating engine performance, for example, the first thing he does is to get all of the facts. He takes nothing for granted, but asks questions of the pilot to determine such things as: Did the trouble come about suddenly, or had you noticed a gradual decrease in performance? Under what conditions of altitude, humidity, temperature, or power setting does this performance loss show up? Does temporarily switching to a single magneto

cause any change in performance? What effect did leaning the mixture, or applying carburetor heat have on the problem? Did switching from one fuel tank to another, or turning on the fuel boost pump have any effect on the problem?

After getting all of the facts, the next step is to eliminate all of the areas that are not likely to cause the trouble. For example: if the magneto drop is normal, but there is a loss of power, the ignition system is more than likely not the problem.

The troubles on the following pages, with some suggested causes, are definitely not all-inclusive, but they do give an idea of some of the most generally found troubles and some of the areas to examine first:

ENGINE IS HARD TO START	
Problem	*Correction*
Too much fuel, the engine is flooded	Place mixture control in cut-off, open the throttle, and keep cranking.
Too much air	Close throttle to about 114″ from full-closed.
Impulse coupling not operating	Turn engine slowly *with switch off*, and listen for impulse coupling snapping.
Fouled spark plug or defective ignition harness	Remove and check spark plugs and check the leads with a harness tester.
Inoperative starting vibrator	Check for proper voltage and listen for vibrator buzzing.
Improper magneto timing	Check both the internal timing of the magneto and the timing of the magneto to the engine.

ENGINE IDLES ROUGH	
Problem	*Problem*
Idle mixture too rich or too lean	Adjust idling mixture and idling RPM.
Plugged injector nozzles	Remove and clean the nozzles in acetone or MEK.
Induction air leak	Check entire induction system and listen for whistling sound when the engine is idling.
Cracked engine mount or defective shock mounts	Repair engine mount or replace shock mounts.
Improperly adjusted fuel pressure	Adjust it to the engine manufacturer's specifications.
Uneven cylinder compression	Perform a differential compression check and remedy the cause of low compression.
Fouled spark plug or defective ignition harness	Remove and check spark plugs and check the leads with a harness tester.

ENGINE WILL NOT DEVELOP FULL STATIC RPM

Problem	Correction
Restricted air inlet	Clean or replace air filter. Check full travel of carburetor heat valve.
Propeller out of adjustment	Check low-pitch angle and adjust if necessary.
Governor out of adjustment	Adjust the governor *after* making very sure that everything else is working properly first.
Internal blockage of muffler	Check the muffler for broken baffles and replace if any are found.

ENGINE WILL NOT DEVELOP RATED POWER

Problem	Correction
Induction system leak	Check entire induction system, and listen for whistling sound when the engine is idling.
Exhaust system leak on turbocharged engine	Check for presence of *feather* indications around exhaust system
Restricted fuel-flow	Check all filters in fuel system. Check for full opening of fuel selector valve. Check calibration of fuel-flow gage.
Restricted air inlet	Clean or replace air filter. Check full travel of carburetor heat valve.
Improper grade of fuel	Check color of fuel to be sure it conforms to that recommended for engine.
Carburetor controls not properly adjusted	Check rigging on all controls to be sure all stops are reached.
Internal blockage of muffler	Check the muffler for broken baffles and replace if necessary.
Low compression	Perform a differential compression check and a borescope inspection if it is needed.

LOW ENGINE OIL PRESSURE

Problem	Correction
Insufficient oil	Carry the correct grade and amount of oil specified by the aircraft manufacturer.
Pressure relief valve out of adjustment	Adjust to engine manufacturer's specifications.
Dirt or carbon under oil pressure relief valve	Remove valve and clean it.
High oil temperature	Check oil grade and quantity. Check oil cooler for air obstructions. Check thermostatic oil temperature control valve. Check for excessive blow-by past piston rings.
Oil pump inlet restricted	Clean oil pick-up screen.

HIGH OIL CONSUMPTION

Problem	Correction
Improper grade of oil	Use only grade specified in the aircraft service manual.
New rings not properly seated	Use only straight mineral oil for break-in period unless otherwise specified by the engine manufacturer. Be sure that break-in is conducted in strict accordance with engine manufacturer's recommendations.
Worn or damaged piston rings or cylinder walls	Perform a differential compression test, and if this indicates bad rings, pull a cylinder and verify with a dimensional check.
Worn valve guides	Remove cylinder and dimensionally check the valve-to-guide clearance. Replace the guide if it is excessive.

HIGH CYLINDER HEAD TEMPERATURE

Problem	Correction
Cooling baffles missing or broken	Check all baffles and be sure they conform to those shown in the parts manual.
Partially plugged fuel injection nozzles	Remove nozzles and clean them in acetone or MEK.
Induction system leak	Check entire induction system and listen for whistling sound when engine is idling.

Glossary

This glossary of terms is provided to serve as a ready reference to the words with which you may not be familiar. These definitions may differ from those of standard dictionaries but are in keeping with shop usage.

black light Ultraviolet light whose rays are in the lower end of the visible spectrum. While more or less invisible to the human eye, they excite or make visible such materials as fluorescent dyes.

brake specific fuel consumption (BSFC) Number of pounds of fuel burned per hour to produce one brake horsepower.

British thermal unit (Btu) The amount of heat required to raise the temperature of one pound of water one degree Fahrenheit.

caliper, micrometer A precision measuring device having a single movable jaw, advanced by a screw. One revolution of the screw advances the movable jaw 0.025″.

caliper, vernier micrometer A micrometer caliper with a special vernier scale which allows each one-thousandth of an inch to be broken down into ten equal parts so one ten-thousandth of an inch may be accurately read.

cam-ground piston An aircraft engine piston ground in such a way that its diameter parallel to the wrist pin boss is less than its diameter perpendicular to the boss. When the piston reaches its operating temperature, the difference in mass has caused the piston to expand to a perfect circular form.

choke bore A method of boring a cylinder of an aircraft engine in which the top, that portion affected by the mass of the cylinder head, has a diameter slightly less than that of the main bore of the barrel. When the cylinder reaches operating temperature, the mass of the head has caused the bore to expand so it is straight throughout its length.

combustion chamber That portion of the cylinder of a reciprocating engine in which the combustion actually takes place. It is that portion above the piston.

compensated relief valve An oil pressure relief valve with a thermostatic valve to decrease the regulated pressure when the oil warms up. High pressure is allowed to force the cold oil through the engine, but the pressure automatically decreases when the oil warms up.

compression ratio The ratio of the volume of an engine cylinder with the piston at bottom center to the volume with the piston at top center.

compression ring The top piston ring used to provide a seal for the gases in the cylinder, and to transfer heat from the piston into the cylinder walls.

crocus cloth An abrasive cloth having a very fine, dark red abrasive on its surface. It is used for polishing metals.

cylinder baffles Thin sheet metal covers and deflectors attached to air-cooled cylinders to force air through the cooling fins to remove the maximum amount of heat.

detonation The almost instantaneous release of heat energy from fuel in an aircraft engine caused by the fuel-air mixture reaching its critical pressure and temperature. It is an explosion rather than a smooth burning process.

dynamic damper A counterweight on a crankshaft of an aircraft engine. It is attached in such a way that it can rock back and forth while the shaft is spinning, and absorb dynamic vibrations. It, in essence, changes the resonant frequency of the engine-propeller combination.

efficiency, mechanical The ratio of the brake horsepower produced by an engine to its indicated horsepower.

efficiency, thermal The ratio of the amount of heat energy converted into useful work to the amount of heat energy in the fuel used.

efficiency, volumetric The ratio of the volume of the charge taken into a cylinder, reduced to standard conditions, to the actual volume of the cylinder.

engine, aircraft An engine that is used to propel an aircraft. It includes the turbochargers and accessories necessary for its functioning, but does not include the propeller.

engine, dry sump An engine in which most of the lubricating oil is carried in an external tank and is fed to the pressure pump by gravity. After it has lubricated the engine, it is pumped back into the tank by an engine-driven scavenger pump.

engine, external combustion A form of heat engine in which the chemical energy in the fuel is converted into heat energy outside of the engine.

engine, four-cycle The most common event cycle for aircraft engines. The four-stroke, five-event cycle consists of an intake stroke, in which the piston moves inward with the intake valve open, and a compression stroke in which the piston moves outward with both valves closed. Near the top of the compression stroke, ignition occurs. The power stroke is an inward stroke of the piston with both valves closed, and the exhaust stroke occurs when the piston moves outward with the exhaust valve open. At this point, the cycle begins again.

engine, gas turbine A form of heat engine in which the burning fuel adds energy to the compressed air and accelerate the air through the remainder of the engine. Some of the energy is extracted to turn the air compressor and the remainder accelerates the air to produce thrust. Some of this energy can be converted into torque to drive a propeller or a system of rotors for a helicopter.

engine, horizontally opposed An engine with cylinders lying flat in two rows, one on either side of the crankcase.

engine, in-line An engine with all of the cylinders in a single line. The crankcase may be located 96 either above or below the cylinders. If it is above, it is called an inverted in-line engine.

engine, internal combustion A form of heat engine in which the chemical energy in the fuel is converted into heat energy inside the engine.

engine, reciprocating An engine which converts the heat energy from burning fuel into the reciprocating movement of the pistons. This movement is converted into rotary motion by the connecting rods and crankshaft.

engine, remanufactured An engine assembled by the engine manufacturer or his authorized agent built up of used parts which are held to the new parts dimensional limits. The engine is given zero time records, and usually the same warranty and guarantee as a new engine.

engine, rotary radial A form of aircraft engine, popular in World War I, in which the propeller was attached to the crankcase and the pistons were attached to an offset cam mounted on the airframe. As the engine runs, the cylinders, crankcase, and propeller all spin around.

engine, static radial An engine with all of the cylinders radiating out from a small central crankcase. A single-throw crankshaft is used for each row of cylinders. All single-row radial engines have an odd number of cylinders, but two or four rows may be used if more power is required.

engine, turboprop A turbine engine which drives a propeller through a reduction gearing arrangement. Much of the energy in the exhaust gases is converted into torque rather than using its acceleration to move the aircraft.

engine, two-cycle A reciprocating engine in which a power impulse occurs on each stroke of the piston. As the piston moves outward, fuel-air mixture is drawn into the crankcase below the piston while above the piston the mixture is compressed. Near the top of the stroke, ignition occurs and, as the piston moves downward, power is exerted on the crankshaft. Near the bottom of the stroke, exhaust action takes place on one side of the cylinder, and intake action occurs on the opposite side.

engine, V- An engine with cylinders arranged in two rows, attached to the crankcase in the form of a "V", with an angle of between 45° and 60° between the banks.

engine, wet sump An engine in which all of the oil supply is carried within the engine itself.

flash point The temperature to which a fluid must be raised before it will momentarily flash, but not sustain combustion when a small flame is passed above its surface.

flight manual Approved information which must be carried in an airplane. This includes the engine operating limits and any other information that is vital to the pilot.

flock Pulverized wool or cotton fibers attached to screen wire used as an air filter. The flock-covered screen is lightly oiled, and it holds dirt and dust, preventing it from entering the engine.

hopper A container within an oil tank used to hold diluted oil for cold weather starting. The use of a hopper minimizes the amount of oil that must be diluted.

horsepower (HP) A practical measurement of power used for aircraft engines. It is the accomplishment of 33,000 ft.-lbs. of work in one minute.

horsepower, brake (BHP) The actual horsepower delivered to the propeller shaft of an engine.

horsepower, friction (FHP) The amount of horsepower required to turn the engine against the friction of the moving parts and to compress the charges in the cylinders.

horsepower, indicated (IHP) The total horsepower developed in the engine. It is the sum of the brake horsepower delivered to the propeller shaft and the friction horsepower required to drive the engine.

interference fit A fit between two parts in which the part being put into a hole is larger than the hole itself. To assemble the parts, the hole is expanded by heating, and the part is shrunk by chilling.

knuckle pin The hardened steel pin that holds an articulating rod in the master rod of a radial engine.

magneto A self-contained, permanent-magnet AC generator with a set of current interrupter contacts and a step-up transformer. It is used to supply the high voltage required for ignition in an aircraft engine.

nitriding A form of case hardening in which the steel part is heated in an atmosphere of ammonia. The ammonia breaks down and its nitrogen combines with aluminum in the steel to form an extremely hard, abrasive-resistant aluminum-nitride surface. Cylinder walls and crankshaft journals are nitrided.

oil, ashless dispersant A popular mineral oil which contains no ash-forming additives, but does contain additives which prevent contaminants clustering together. It keeps the contaminants dispersed throughout the oil.

oil, detergent A mineral oil to which ash-forming additives have been added to increase its resistance to oxidation. Because of its tendency to loosen carbon deposits, it is not used in aircraft engines.

oil, synthetic A lubricating oil with a synthetic rather than petroleum base. It has less tendency toward oxidation and sludge formation than petroleum oils. It is extensively used in turbine engines and is gaining popularity in reciprocating engines.

oil control ring The piston ring below the compression rings used to control the amount of oil between the piston and the cylinder wall. It is usually a multi-piece ring and normally fits into a groove with holes to drain part of the oil back to the inside of the piston.

oil scraper (wiper) ring A piston ring located at the bottom, or skirt end, of a piston used to wipe the oil either toward or away from the oil control ring, depending on the design of the engine.

overhaul, major The complete disassembly, cleaning, inspection, repair, and reassembly of an engine or other component of an aircraft.

overhaul, top The overhaul of the cylinders of an aircraft engine. It consists of grinding the valves, replacing the piston rings, and doing anything else necessary to restore the cylinders to their proper condition. The crankcase of the engine is not opened.

permamold crankcase An engine crankcase which has been pressure-molded in a permanent mold rather than being sand-cast. It is thinner and more dense than a sand-cast crankcase.

petroleum A substance containing a form of chemical energy used as a fuel for most of our aircraft engines. It is a natural hydrocarbon product which was, in ancient times, plant or animal life but was buried under billions of tons of earth. It is obtained as a liquid from deep wells.

piston displacement The total volume swept by the pistons of an engine in one revolution of the crankshaft.

planetary gears A reduction gearing arrangement in which the propeller shaft is attached to an adapter holding several small planetary gears. These gears run between a sun gear and a ring gear, either of which may be driven by the crankshaft, and the other is fixed into the nose section. Planetary gears are efficient and do not reverse the direction of rotation between the two shafts.

porous chrome plating An electrolytically deposited coating of chromium on walls of aircraft engine cylinders. The surface contains thousands of tiny cracks which hold oil to provide for cylinder wall lubrication.

pour point The lowest temperature at which a fluid will pour without disturbance.

power (P) The time rate of doing work. It is force times distance divided by time.

preignition   Ignition occurring in the cylinder before the time of normal ignition. It is often caused by a local hot spot in the combustion chamber igniting the fuel-air mixture.

pressure, brake mean effective (BMEP)   A computed value (not measured) of the average pressure that exists in the cylinder of an engine during the power stroke.

pressure, indicated mean effective (IMEP)   The average, actually measured pressure in the cylinder of an engine during the power stroke.

pressure, manifold   The absolute pressure measured at the appropriate point in the induction system of an aircraft engine, and usually expressed in inches of mercury.

preventive maintenance   Simple or minor preservation operations and replacement of small standard parts not involving complex assembly operations.

propeller   A device for propelling an aircraft that, when rotating, produces by its action on the air, a thrust approximately perpendicular to its plane of rotation. It includes components normally supplied by its manufacturer.

pump, gear   A form of constant displacement pump in which two spur-type gears mesh and rotate within a close fitting housing. Oil is picked up as the teeth come out of mesh, is carried around the outside of the gears, and discharged from the pump when the teeth mesh.

pump, gerotor   A form of constant displacement pump using a spur gear driven by the engine and turning inside an internal tooth gear having one more space than teeth on the drive gear. As the pump rotates, the volume at the inlet port increases, and that at the outlet decreases, moving fluid through the pump.

pump, scavenger   A constant displacement pump in an engine that picks up oil after it has passed through the engine and returns it to the oil reservoir.

quill shaft   A hardened steel shaft with round cross section and splines on each end. Torsional flexing of the shaft is used to absorb torsional vibrations.

rated maximum continuous power   The maximum approved brake horsepower developed by an aircraft engine approved for an unrestricted period of time.

rated takeoff horsepower   The approved brake horsepower that is allowed to be developed by an aircraft engine for a period not exceeding five minutes.

resonant   A condition in which a mechanical system is allowed to vibrate when its natural frequency is exactly the same as the frequency of the applied force.

rod, articulating   The rod in a radial engine that connects the piston to the master rod. It rocks back and forth rather than encircle the crankshaft. It is also called a link rod.

rod, master   The only connecting rod in a radial engine whose big end passes around the crankshaft. All of the other rods connect to the master rod and oscillate back and forth rather than encircling the crankshaft.

solenoid   A coil of wire with a movable core. In magnetic particle inspection, it is the coil used to longitudinally magnetize the part and to demagnetize parts using AC.

specific fuel consumption   The number of pounds of fuel consumed in one hour to produce one horsepower.

spectrometric oil analysis   A system of oil analysis in which a sample is burned in an arc and the resulting light is examined for its wavelengths. This test can determine the amount of the different metals suspended in the oil, and can give an indication of an impending engine failure.

stellite   An extremely hard and wear-resistant metal used for valve faces and stem tips. It contains cobalt, tungsten, chromium, and molybdenum.

tachometer   An instrument which measures the rotational speed of an engine in revolutions per minute (RPM).

Time Between Overhauls (TBO)   A recommendation of the manufacturer of an aircraft engine as to the amount of time that the engine can operate under average conditions before it should be overhauled. Overhauls at this time will result in the most economical operation.

turbocharger   An air compressor used to increase the pressure of the air entering the fuel metering system. The compressor is driven by a high-speed turbine which is spun by the exhaust gases leaving the engine.

valve clearance, cold   The clearance between the valve stem and the rocker arm of an engine using solid valve lifters when the engine is cold.

valve clearance, hot   The clearance between the valve stem and the rocker arm of an engine using solid valve lifters when the engine is at operating temperature.

valve clearance, timing The clearance to which a poppet valve using solid lifters is adjusted to set the cam-to-crankshaft timing. The valves in one cylinder are adjusted to this clearance, the timing is set, and the valves are then re-adjusted to the cold clearance.

valve float A condition in which the frequency of the valve opening exactly corresponds to the resonant frequency of the valve spring. Under these conditions, the valve spring will exert no closing force.

valve lag The number of degrees of crankshaft rotation after top or bottom center at which the intake or exhaust valve opens or closes. For example, if the intake valve closes 60° after bottom center on the compression stroke it has a valve lag of 60°.

valve lead The number of degrees of rotation before top or bottom center at which the intake or exhaust valve opens or closes. For example, if an intake valve opens 15° before the piston reaches top center on the exhaust stroke, it is said to have a 15° valve lead.

valve overlap The angular distance of crankshaft rotation when the piston is passing top center on the exhaust stroke when the intake and exhaust valves are both open.

varsol A petroleum product similar to naphtha, used as a solvent for washing aircraft engine parts.

venturi A specially shaped restrictor in a fluid-flow passage used to increase the velocity of the fluid and decrease its pressure.

Vibratory Torque Control (VTC) A special patented coupling between the crankshaft and the propeller shaft of a Continental Tiara engine. It incorporates a quill shaft to absorb torsional vibrations and a centrifugally actuated lock to lock out the quill shaft for operations when a solid shaft would be more advantageous.

viscosity The fluid friction, or the resistance to flow, of an oil or other fluid.

viscosity index The measure of change in viscosity of an oil with a change in temperature.

work The product of force and distance.

wrist pin A hardened and polished steel pin that attaches the small end of a connecting rod into a piston.

Y-valve The oil drain valve for a dry sump engine. It derives its name from its shape. One arm of the Y goes to the pressure pump inlet of the engine, one arm to the oil reservoir, and the lower arm is fitted with a shut-off valve, and is the point from which the oil may be drained from the reservoir. Fuel for oil dilution is introduced in the Y-valve.

Answers To Study Questions

Chapter 1

1. A Btu, or British Thermal Unit, is the amount of heat energy required to raise the temperature of one pound of pure water 1°F.

2. Steam-driven turbine

3. A rocket engine uses a fuel which contains its own oxygen and is therefore not dependent on the oxygen in the air.

4. The two-stroke cycle engine is too inefficient and difficult to cool for it to gain popularity as an aircraft engine.

5. a. About 15° before top center at the end of the exhaust stroke
   b. About 60° after the piston has passed bottom center at the beginning of the compression stroke
   c. About 60° before the piston reaches bottom center on the power stroke
   d. About 15° after top center at the beginning of the intake stroke
   e. About 30° before the piston reaches top center on the compression stroke.

6. Work is the product of force times distance, and power takes into consideration the time required to do the work.

7. 542.5 watts

8. 73 BHP

9. 30%

10. a. The bore
    b. The stroke
    c. The number of cylinders

11. The anti-detonation characteristics of the fuel used

12. If the ignition occurs too late, there will not be sufficient time for the gases to be completely burned before the exhaust valve opens. Burning gases flowing over the exhaust valve will cause overheating.

13. The RPM should be increased, spreading the energy release over a greater number of power strokes.

Chapter 2

1. When an engine is operated for only an hour or so a week, the moisture which condenses in the crankcase does not have an opportunity to evaporate, and it forms acid contaminants in the oil that will decrease the engine's service life.

2. An in-line engine requires such a long crankshaft that its weight for its horsepower is less desirable than that of other configurations.

3. The V-engine is more compact and weighs less than an in-line engine, and it has far less frontal area than a radial engine.

4. Its low weight for its power

5. Its low weight for its power

6. a. It has a small frontal area
   b. It has a good weight-for-power ratio

7. Ethylene glycol

8. Ram air into the engine cowling creates an area of high pressure above the engine, and low pressure below. The only way cooling air can pass from one area to the other is through the cylinder fins. In passing through, it removes heat from the cylinder.

9. A wet sump engine is one in which the lubricating oil is carried in a sump which is part of the engine itself.

10. Number 1 cylinder is the top cylinder in the rear row.

11. Number six-left (6L)

12. The front cylinder on the right side

13. The rear cylinder on the right side

14. 1-3-5-7-9-2-4-6-8

15. 1-12-5-16-9-2-13-6-17-10-3-14-7-18-11-4-15-8

16. 1-3-2-4. For counterclockwise rotation, it is 1-4-2-3.

17. 1-6-3-2-5-4

18. High-strength chrome molybdenum steel

19. The head is heated and screwed onto the threaded barrel. This forms a gas-tight interference fit.

20. a. The microscopic cracks in the chromed surface hold oil which provides good lubrication and prevents rust.
    b. The hard chrome walls do not wear as much as the softer steel walls.

21. The cylinder barrel has been chrome plated.

22. Nitriding is a form of case hardening in which nitrogen from ammonia gas combines with aluminum in the steel to form a hard, wear-resistant aluminum nitride surface on the steel.

23. Nitrided surfaces are susceptible to pitting from moisture in the air.

24. The hole in which the guide fits is smaller than the guide. The cylinder head is heated, expanding the hole. The guide is chilled, shrinking it. The two are then pressed together.

25. Metallic sodium melts when the engine is operating and sloshes back and forth in the valve; thus it absorbs heat from the valve 102 head and transfers it through the valve guide into the cylinder head where it can be dissipated into the air through the cylinder fins.

25. Forged aluminum alloy

27. These are the compression rings and are used to seal the gases in the cylinder and transfer heat to the cylinder walls.

28. This is the oil control ring used to control the oil film on the cylinder wall.

29. This is the oil wiper ring and is used to wipe oil either toward or away from the oil control ring, depending on the design of the engine.

30. Plain gray cast iron

31. They are full-floating and are free to rotate in both the piston and the connecting rod.

32. Only one rod, the master rod, attaches directly between the crankshaft and the piston; all of the other pistons are connected by link rods to the master rod.

33. 60°

34. They are case hardened by nitriding

35. Dynamic dampers change the resonant frequency of the crankshaft-propeller combination to prevent its being excited by the firing impulses of the engine.

36. A reduction gearing arrangement is needed to allow the engine to develop its desired RPM while holding the tip speed of the propeller below the speed of sound.

37. A quill shaft flexes torsionally and absorbs some of the torsional vibrations that could damage the engine.

38. 1.5:1

39. The hot, or running, clearance is much greater than the cold clearance.

40. A hydraulic valve lifter maintains zero clearance between the rocker arm and the valve stem, minimizing noise and wear.

41. Cast aluminum alloy

42. A circular, tapered seal held against the shaft by a coil or a flat-leaf-type spring

43. Straight mineral oil has a tendency to oxidize when it is exposed to elevated temperatures or when it is aerated.

44. Detergent oil loosens carbon deposits which have formed in the engine, and these loosened deposits are likely to plug oil passages and filters.

45. Additives that cause the sludge-forming materials in the oil to disperse and remain in suspension rather than joining together.

46. No, all of the brands of AD oil are compatible. There should be no problem from mixing them.

47. SAE 40

48. Oil for the valve operating mechanism flows into the rocker boxes through hollow push rods.

49. A thermostatic valve causes cold oil to bypass the oil cooler.

50. A gerotor pump is a constant displacement pump.

51. A compensated oil pressure relief valve allows the oil pressure to be high enough to force cold oil through all of the passages, yet lowers the pressure when the oil heats up.

52. Oil dilution thins the oil before shutdown when an extremely cold start is anticipated the next morning.

53. Most engines are equipped with a full-flow filter.

54. The non-uniformity of the mat may allow high-pressure oil to channel through and lower the efficiency of the filter.

55. The can in which the filter element is sealed is cut open, and the element is taken apart and examined.

56. A small restricting orifice is installed between the engine and the gage.

57. Oil temperature is taken as the oil enters the engine.

58. These baffles direct cooling air between the fins on the cylinders to remove heat.

59. Varying the amount of opening of the cowl flaps varies the pressure differential between the top and bottom of the engine and controls the amount of cooling airflow through the cylinders.

60. Augmenter tubes lower the pressure below the engine, increasing the airflow through the fins.

61. a. Weigh the amount of air entering the engine.
    b. Meter into this air the proper amount of fuel.
    c. Mix the fuel and air.
    d. Distribute the fuel-air mixture to all of the cylinders.

62. The atmospheric pressure in the float bowl

63. The main metering jet

64. The airflow into the engine creates a pressure difference across a diaphragm which is the difference between impact air and venturi air pressure. This measures the amount of air entering the engine. The movement of the diaphragm opens a poppet valve to admit exactly the proper amount of fuel for the airflow.

65. It tends to enrich.

66. The throttle stop which controls the amount of air allowed to flow past the throttle valve.

67. It provides a temporarily rich mixture during the transition from the idle system to the main metering system.

68. The power enrichment system gives the engine an excessively rich mixture for full power operation to remove some of the heat from the cylinders.

69. The main difference is the location at which the fuel is actually introduced into the engine. With a carburetor, it is at the beginning of the induction system. With a fuel injection system, it is in the cylinder heads.

70. In a fuel injection system, only the air flows through the induction pipes which have different lengths. A uniform amount of fuel is discharged through equal length injector lines directly into each of the cylinders.

71. a. Flocked screen wire
    b. Disposable paper element
    c. Glycol-impregnated polyurethane foam element

72. They are washed in a mild soap and water solution and allowed to dry.

73. Heated induction air at high power settings can cause detonation.

74. The air drawn into the engine when the carburetor heat valve is in the hot position is not filtered, and abrasive dust can be ingested into the cylinders.

75. The existing atmospheric pressure

76. An engine with a single-speed blower whose takeoff horsepower is boosted above that which the engine could develop with normal aspiration.

77. A small turbine driven by the exhaust gases

78. By varying the amount of exhaust gases that are allowed to flow through the turbine.

79. The manifold pressure

80. Corrosion resistant alloy steel

81. It may be used to warm the cabin or to prevent carburetor ice.

82. Back pressure opposes the flow of exhaust gases from the cylinders and reduces the power output of the engine.

83. The PRT extracts some of the energy from the exhaust gases and returns it to the crankshaft as mechanical energy.

84. About 30° before the piston reaches top center on the compression stroke

85. a. For safety in case one of the magnetos should fail
    b. To assure more even combustion by igniting the mixture in two places in the cylinder

86. Magneto ignition is completely self-contained and is independent of the electrical system of the aircraft.

87. A magneto is a small permanent-magnet AC generator with a set of breaker points to interrupt the current, and a step-up transformer to increase the voltage to a value high enough to jump the gap in the spark plug.

88. The spark is produced when the points open, interrupting the flow of primary current.

89. a. It decreases the arcing as the points begin to open.
    b. It assists the flow reversal each time the points close.

90. The points must be adjusted so they just begin to open at the instant the rotating magnet is in its E-gap position.

91. The number of degrees beyond the neutral position of the rotating magnet when the flow of primary current is the greatest.

92. The spark plug conducts high voltage electrical energy into the combustion chamber of the engine where it can jump a carefully controlled gap to produce a spark.

93. Fine-wire spark plugs are less susceptible to fouling from lead deposits from the fuel.

94. This resistor limits the duration of the spark and minimizes electrode erosion.

95. This braid is called shielding and directs radiated electrical energy to ground, preventing radio interference.

96. A propeller converts the torque of the engine into thrust.

97. Low-pitch angle

98. Centrifugal twisting moment or the centrifugal force acting on counterweights opposed by the force of oil pressure acting on a piston.

Chapter 3

1. No. This is a recommendation only and is based on all of the operating conditions being average.

2. A TBO is based on the wear history of an engine. If an engine is overhauled at this time, the parts should not be worn to such a point that major components will need to be replaced.

3. The overhaul of the cylinders without getting into the crankcase or any of the rotating mechanism. It includes grinding and seating the valves, replacing the piston rings, and doing anything to the cylinders to restore them to their original operational condition.

4. A top overhaul is economical if there is no obvious damage or performance deterioration caused by the condition of the cylinders when there is an appreciable amount of time left before the recommended TBO.

5. A careful check should be kept of the trends in compression pressure and oil consumption, and the owner should be informed of the dangers of operating some of the accessories beyond the overhaul times recommended.

6. An excess of tetraethyl lead in the fuel can cause valve sticking, oil contamination, and spark plug fouling.

7. Corrosion

8. Moisture will condense inside the engine and mix with the oil. If the oil is not warmed up thoroughly, more than is possible with a ground run, the water in the oil will form acids which will damage the metal parts of the engine.

9. The inside of all of the cylinders should be sprayed with a preservative oil and the engine not turned until it is depreserved.

10. a. Check the oil quantity
    b. Check for indications of oil leaks
    c. Check for any indication of exhaust leaks and the security of the exhaust system.
    d. Drain the fuel strainer and check for contamination or water.
    e. Check the ignition system for proper magneto drop and for the proper operation of the ignition switch.
    f. Check the propeller pitch change mechanism for proper cycling.
    g. Check for proper static RPM and manifold pressure.

11. It is not required, but is good operating practice.

12. An A&P mechanic holding an Inspection Authorization

13. a. A direct compression test
    b. A differential compression test

14. The exhaust valve in the cylinder being tested is leaking.

15. It should be cut open and the element examined for contamination.

16. This is an indication of excessive oil consumption in that cylinder, most probably caused by worn or broken piston rings, or excessively worn valve guides.

17. Hotter spark plugs should be installed.

18. Trichlorethylene

19. A leaking exhaust muffler can allow deadly carbon monoxide to enter the aircraft cabin.

20. No. Because of the alternating voltage produced in a magneto, the spark plugs should be returned to the next cylinder in firing order to even out the erosion between the center and the ground electrodes.

21. a. New parts limits
    b. Serviceable limits

22. Only the manufacturer of the engine or a repair station approved by the engine manufacturer.

Chapter 4

1. Usually the magnetos, the carburetor or fuel injection system, the induction system between the carburetor and the cylinders, and all of the intercylinder baffles.

2. The alkali in this type of compound can remain in the pores of the metal and after the engine is returned to service will mix with the oil to form soap. This causes foaming and can lead to failure of the lubrication system.

3. An approved decarbonizing solution

4. Binder twine may be pulled through the ring grooves.

5. It should be vapor degreased with hot trichlorethylene vapors.

6. a. Circular
   b. Longitudinal

7. The large clearances which exist when serviceable limits are used will cause accelerated wear and require more extensive replacement of parts when the engine is next overhauled.

8. After the crankcase halves are assembled and properly torqued with new bearing inserts installed, the inside diameter of the bearing is measured with a telescoping gage and a micrometer, and the outside diameter of the bearing journal is measured with the same micrometer. The difference between the two readings is the clearance.

9. An interference fit is one in which the hole is smaller than the part which fits into it.

10. The engine manufacturer or repair station approved by the FAA for this specialized type of work.

11. The camshaft bearings may be line-bored oversize and a camshaft with oversized bearings installed.

12. The stud is 0.003" oversize on the coarse threaded end.

13. It must be renitrided.

14. Its inside surface may be built back up by porous chrome plating and then ground to proper size.

15. a. The seat is ground with a stone of the proper angle.

    b. A 15° stone is used to get the proper diameter of the top of the seat.

    c. A 75° stone is used to narrow the seat to the proper width by removing metal from the lower part of the seat.

16. Only as much as is necessary to remove any pits or wear marks.

17. The valve

18. The spring is magnetically inspected for cracks, and then the amount of force required to compress it to a specific height is measured.

19. The ring is put inside the wear area of the cylinder and squared up by pushing against it with the piston head. When the ring is in the proper position, the gap is measured with a feeler gage.

20. The ring is held flush with the outside of the piston by forcing it out against a straight edge on the piston surface. The space between the ring and the side of the groove is measured with a feeler gage.

21. A plain cast-iron ring

22. a. The diameter of the small end bushing

    b. The diameter of the big end bearing

    c. The amount of twist in the rod

    d. The amount of bend in the rod

23. It is almost impossible to completely demagnetize the ball, and even the slightest amount of magnetization will prevent the ball seating.

24. The lifter assembly is cleaned, dried, and assembled. The plunger is quickly depressed into the cylinder, and if the assembly is working properly, it will bounce back.

25. They are rolled across a surface plate.

26. A very thin coating of an approved non-hardening gasket compound is spread on both mating surfaces and a very fine silk thread is embedded into this compound.

27. The sequence in which the fasteners are torqued down

28. a. By the use of a dynamometer

    b. By the use of a special test club propeller

29. These instruments do not normally have sufficient accuracy.

30. a. Use a pressure pre-oiler to force oil through all of the passages.

    b. Remove one spark plug from each cylinder and spin the engine with the starter until there is an indication of oil pressure on the oil pressure gage.

31. After the initial run-in period, the oil is drained, weighed, and then put back into the engine. The oil consumption run is made after which the oil is again drained and weighed.

32 Normal power for climbs and 75% power for cruise

33. Only straight mineral oil

Final Examination

Place a circle around the letter for the correct answer.

1. **Where is the piston located when the intake valve opens?**
   A. Before the piston reaches the top of the exhaust stroke
   B. Before the piston reaches the bottom of the intake stroke
   C. Before the piston reaches the top of the compression stroke
   D. Before the piston reaches the bottom of the power stroke

2. **How much aviation gasoline will an engine burn if it produces 180 brake horsepower and has a brake specific fuel consumption of 0.47?**
   A. 84.6 gallons per hour
   B. 14.1 gallons per hour
   C. 14.1 pounds per hour
   D. 42.3 pounds per hour

3. **With what material are exhaust valves partially filled to aid in transferring heat from?**
   A. Sodium chloride
   B. Liquid potassium
   C. Stellite
   D. Metallic sodium

4. **What type of crankshaft is used by a 6-cylinder horizontally opposed engine?**
   A. 45°
   B. 60°
   C. 120°
   D. 180°

5. **Which statement is true regarding a choke-ground cylinder?**
   A. The diameter of the head end of the cylinder is greater than the diameter in the center of the stroke.
   B. The diameter at the skirt end of the cylinder is greater than that in the center of the stroke.
   C. The diameter at the head of the cylinder is smaller than that in the center of the stroke.
   D. The diameter at the skirt end of the cylinder is smaller than that in the center of the stroke.

6. **What should be done to an engine if the spark plugs removed from one of the cylinders are exceptionally clean, while all of the other show normal operational deposits?**
   A. Clean all of the other spark plugs so they will all be alike.
   B. Replace these spark plugs because they have most probably not been firing.
   C. Carefully check the magneto to find why these spark plugs were not firing.
   D. Examine the inside of the cylinder from which these spark plugs were removed for indications of damage from detonation.

7. **Which type of oil is normally recommended for engine break-in?**
   A. Ashless dispersant oil
   B. Straight mineral oil
   C. Detergent oil
   D. Synthetic oil

8. **Which statement is true regarding the manufacturer's recommended TBO?**
   A. The FAA requires that all aircraft engines be overhauled at their recommended TBO.
   B. The manufacturer's warranty is normally voided if the engine is operated beyond the recommended TBO.
   C. Overhauling an engine at the recommended TBO will normally result in the most economical operation, as it will prevent excessive wear of the parts.
   D. The recommended TBO is actually a sales gimmick, and really has little to do with practical operation.

9. **What may be done to an air-cooled cylinder if the barrel is worn slightly out of tolerance but is otherwise in good condition?**
   A. If it is only slightly out of tolerance, go ahead and use it.
   B. It may be ground out, and a steel sleeve pressed in place.
   C. It may be ground smooth, built back up with porous chrome plating, and then ground to size.
   D. It may be built up by nitriding and ground back to size.

10. **If an exhaust valve is required to fit its seat with a line contact, or an interference fit, and the seat is ground to an angle of 45°, the valve will most likely be ground to an angle of**

   A. 44.5°

   B. 45.5°

   C. 45°

   D. 30°

11. **What is not normally done on a top overhaul?**

   A. Hydraulic valve lifter bodies magnetically inspected

   B. Valves and valve seats ground

   C. Piston rings replaced

   D. Push rods inspected for straightness

12. **Which type of magnetization would be used to locate a crack that extends across a wrist pin?**

   A. Circular magnetization

   B. Longitudinal magnetization

13. **What is the purpose of a quill shaft in some of the modern engine crankshafts?**

   A. It increases the torsional strength of the crankshaft.

   B. It absorbs some of the torsional vibrations that are transmitted into the crankshaft.

   C. It aids in dynamically balancing the crankshaft.

   D. It is a special extension shaft used to drive some of the accessories.

14. **What drives a turbocharger compressor?**

   A. An exhaust gas-driven turbine

   B. A hydraulic motor driven by engine oil pressure

   C. Exhaust gases impinging on the compressor scroll

   D. It is gear-driven from an accessory drive of the engine.

15. **What is the purpose of an oil dilution system in an aircraft engine?**

   A. It is used to decrease the oil pressure for long flights at high altitude and low temperatures.

   B. It is used to increase the viscosity index of the oil.

   C. It is used to decrease the viscosity of the oil for cold weather starting.

   D. It is used to control the oil flow during the break-in period.

16. **What is the process of nitriding?**

   A. The application of a thin plating of hard material on the surface of a crankshaft.

   B. A treatment in which the surface of the crankshaft is converted into extremely hard nitrides.

   C. A case hardening treatment in which additional carbon is infused into the surface of the crankshaft.

   D. A case hardening process in which some of the carbon is removed from the surface of the crankshaft.

17. **What is meant by a wet sump engine?**

   A. One that uses hydraulic valve lifters.

   B. One that uses water injection for detonation suppression.

   C. One that carries its oil supply in an external oil tank.

   D. One that carries its oil supply in the engine itself.

18. **What is the brake horsepower developed by a 4-cylinder engine with a 4″ bore, a 4″ stroke, a BMEP of 125 PSI, and turning at 2,300 RPM?**

   A. 146 HP

   B. 723 HP

   C. 73 HP

   D. 876 HP

19. **Who may authorize an engine to be returned to service after a major overhaul with zero time records?**

   A. Any licensed A&P mechanic

   B. Any FAA-approved repair station

   C. An A&P mechanic holding an Inspection Authorization

   D. Only the engine manufacturer or a facility approved by him

20. **Which statement is true regarding the repair of aircraft engine crankcases?**

   A. A cracked crankcase cannot be repaired. It must be replaced.

   B. A cracked crankcase must be returned to the engine manufacturer for repair.

   C. A cracked crankcase may be repaired by a repair station approved by the FAA for this specialized work.

   D. After a crankcase has been repaired, it must be inspected by the magnetic particle inspection method.

Answers To Final Examination

1. A		11. A	
2. B		12. B	
3. D		13. B	
4. B		14. A	
5. C		15. C	
6. D		16. B	
7. B		17. D	
8. C		18. C	
9. C		19. D	
10. A		20. C	